Utilizing Aeroponics Techniques for Improved Farming

Chelladurai G.
Department of Botany, St. Joseph's College (Autonomous), Tiruchirappalli, India

IGI Global
Scientific Publishing
Publishing Tomorrow's Research Today

Vice President of Editorial	Melissa Wagner
Managing Editor of Acquisitions	Mikaela Felty
Managing Editor of Book Development	Jocelynn Hessler
Production Manager	Mike Brehm
Cover Design	Phillip Shickler

Published in the United States of America by
IGI Global Scientific Publishing
701 East Chocolate Avenue
Hershey, PA, 17033, USA
Tel: 717-533-8845
Fax: 717-533-8661
Website: https://www.igi-global.com E-mail: cust@igi-global.com

Library of Congress Cataloging-in-Publication Data

Names: Chelladurai, G., DATE- editor.
Title: Utilizing aeroponics techniques for improved farming / edited by Chelladurai G.
Description: Hershey, PA : IGI Global Scientific Publishing, [2025] | Includes bibliographical references and index. | Summary: "The objective of this book is to provide comprehensive insights into the principles, practices, and benefits of utilizing aeroponics techniques in modern agriculture"-- Provided by publisher.
Identifiers: LCCN 2024040531 (print) | LCCN 2024040532 (ebook) | ISBN 9798369323205 (hardcover) | ISBN 9798369350812 (paperback) | ISBN 9798369323212 (ebook)
Subjects: LCSH: Hydroponics. | Alternative agriculture. | Agricultural innovations.
Classification: LCC SB126.5 .U85 2024 (print) | LCC SB126.5 (ebook) | DDC 631.5/85--dc23/eng/20240924
LC record available at https://lccn.loc.gov/2024040531
LC ebook record available at https://lccn.loc.gov/2024040532

British Cataloguing in Publication Data
A Cataloguing in Publication record for this book is available from the British Library.

Table of Contents

Detailed Table of Contents

Growing plants in air or a mist without the use of soil or aggregate media is known as aeroponics. To promote the best possible plant growth, air cultures enhance air exposure. Aeroponically grown plants require materials and devices that are devoid of any disease or infections. Minimal plant support elements are an essential component of a true aeroponic culture and apparatus. By reducing the amount of contact between the plant and the support structure, this design maximizes the airflow surrounding the plant. Ensuring the root systems are unrestricted is essential for long-term aeroponic cultivation because this will allow for unimpeded access to clean water, sufficient air exchange, and disease-free environments. In aeroponics, clean air is essential to the purification of the plant habitat. Air must be freely accessible for plants to grow naturally and for their physiological development to be successful. So, the development of aeroponics, as well as its fundamental ideas and important techniques, are briefly discussed in this chapter.

Finding sustainable solutions is becoming more and more important for the agricultural sector as environmental pollution and climate change worsens. Through interventions of technologies such as ICT, sensor and actuator technology made smart farming easier. But there have been difficulties along the way with smart farming, especially with indoor hydroponic systems. High costs and the inability to compete with the pricing of traditionally grown commodities have forced many businesses to file for bankruptcy. So, to overcome the issues the researchers came up with a renowned idea called Aeroponics. This chapter elaborated the History, Classification, Working, Advantage, Disadvantage along with difference between hydroponic and aeroponic.

Aeroponic means "growing in air." An aeroponic system is medium-less in that the roots of the plant are free-hanging inside an open root-zone atmosphere. Aeroponics structure supplies optimum levels of water, nutrients and air to the growing chamber. Aeroponics is the process of growing plants in an air or mist environment without use of soil or aggregate media. The word aeroponic is derived from the Latin word 'aero' (air) and 'ponic' means labor (work). This is an alternative method of soil-less culture in growth-controlled environments. The aeroponic culture technique is an optional device of soil-less culture in growth-controlled environments such as greenhouses. Aeroponics is more than just a farming method; it represents a paradigm shift in agriculture, offering a sustainable and efficient alternative to traditional farming practices. By embracing innovative techniques like aeroponics, we can work towards a more sustainable and resilient agricultural system that meets the needs of growing global population while preserving our precious water resources

The chapter "Working Method of Aeroponics: The article "Innovations in Soilless Agriculture" explains how aeroponic farming works, and the parts of an aeroponic system, with particular concentration on particular farming procedures. Aeroponics provides an organic technique to address contemporary horticulture's problems by delivering suspended plant roots in air with nutrient enriched mist. Its advantages include better nutrient uptake, good aeration and very reduced water and other resource use as compared to conventional soil-based farming. Each of them is elaborated on in the chapter and described its influence on the growth of plants in the given aeroponic system; these components include the growth chamber, the system of misting the leaves, nutrient supplying, and the environmental factors regulating the plant growth.

This chapter will delve into the integration of aeroponics into artificial intelligence and deep learning techniques for agricultural productivity and its sustainability. This is a no-soil farming method where plants grow in a nutrient-rich mist. Therein lies a couple of major advantages: water efficiency and an accelerated pace of plant growth. The reasoning behind the inclusion of AI and deep learning techniques lies in the computer vision and predictive analytics ability—how best it can help determine if this has the potential to optimally run aeroponic systems. Monitoring the health of a plant and the environmental state in real time, using AI-driven sensors and deep learning for analytics, is capable of identifying data patterns in predicting growth and optimizing practices for the delivery of nutrients. Some specific successful cases and novel innovations are exemplified next, showing how new breakthroughs can solve the existing challenges of agriculture, improve yield quality, and ultimately reduce resource consumption.

It is the air above the land that will shape farming in the future. Modern methods of agriculture such as Aeroponics, which grows plants in an air or mist environment without the need for soil, are reimagining traditional Agriculture in response to growing global issues like climate change, water shortage, and population expansion. This innovative strategy has the power to completely change the way we grow food,

This chapter explores advanced IoT solutions designed to boost agricultural productivity by integrating cutting-edge technologies. It examines the deployment of IoT sensors and devices for real-time monitoring of soil conditions, crop health, and environmental factors. Key innovations discussed include precision irrigation systems, automated climate control, and predictive analytics that leverage big data and machine learning to optimize crop yields. The chapter also highlights case studies demonstrating successful IoT applications in various agricultural settings, emphasizing the impact on resource efficiency and yield improvement. Challenges such as data security, integration with existing systems, and cost implications are addressed, along with strategies for overcoming these obstacles. By providing a comprehensive overview of current IoT technologies and their practical applications, this chapter offers valuable insights for researchers, practitioners, and policymakers aiming to enhance agricultural productivity through technological advancement.

The chapter focuses on artificial intelligence in its integration into modern systems of aquaponics and on its modernizing potential in contemporary agriculture. AI-driven solutions enhance the efficiency, sustainability, and productivity of aquaponics through optimization of water quality, nutrient cycling, and fish-plant interactions. This chapter will look into the applications of AI in machine learning algorithms for predictive analytics, automated monitoring systems for real-time data collection, and decision support systems in dynamic resource management. It also involves scaling the operations of aquaponics through the role AI can play in increasing crop yield while reducing the impact on the environment. Case studies have been done showing successful implementations of AI-enhanced aquaponics, showing system resilience and efficiency in resource use. Finally, the chapter concludes with the indication of future directions for AI in aquaponics as a pathway toward more sustainable and intelligent agricultural practices with environmental sustainability goals.

Aeroponics, a soilless cultivation technique, is gaining momentum as a transformative solution in modern agriculture. This chapter explores its commercial applications, focusing on its ability to revolutionize traditional farming and support sustainable food production. By delivering nutrients directly to plant roots in a controlled environment, aeroponics offers benefits like faster growth rates, higher yields, and reduced resource use. The chapter examines its adoption in urban farming, vertical farming, and greenhouse operations, along with its economic viability and environmental impact. Through case studies, the potential of aeroponics to shape the future of agriculture is highlighted, offering a practical overview of its commercial prospects.

Preface

Agriculture has always been the backbone of human civilization, continuously evolving to meet the growing demands of an expanding population. As challenges such as limited arable land, climate change, and water scarcity threaten traditional farming methods, innovative solutions are required to ensure sustainable food production. Aeroponics, a cutting-edge soilless cultivation technique, offers a promising alternative by maximizing plant growth while minimizing resource consumption. Aeroponics is a revolutionary farming method that suspends plant roots in the air, delivering nutrients and water in a fine mist. Unlike conventional soil-based agriculture or even hydroponics, aeroponics enhances oxygen availability, accelerates biomass production, and optimizes nutrient absorption. Studies, including research conducted by NASA, have demonstrated that aeroponics can reduce water usage by up to 98%, making it an ideal solution for future food security. This technique is particularly valuable in urban environments, where space constraints and resource efficiency are paramount. This book, Utilizing Aeroponics Techniques for Improved Farming, aims to provide a comprehensive exploration of aeroponic farming principles, practices, and benefits. It delves into the scientific foundation of aeroponics, its applications in modern agriculture, and its potential role in addressing global food crises. By offering insights into system design, nutrient management, and environmental control, this book serves as a valuable resource for researchers, farmers, agronomists, and technology enthusiasts interested in adopting and advancing aeroponic techniques.

Chapter 1
Aeroponics System and Its Methods

Muthuminal Ramu
https://orcid.org/0000-0001-9931-2204
SNS College of Technology, India

R. Mohana Priya
SNS College of Technology, India

R. M. Swarna Priya
Vellore Institute of Technology, India

ABSTRACT

Growing plants in air or a mist without the use of soil or aggregate media is known as aeroponics. To promote the best possible plant growth, air cultures enhance air exposure. Aeroponically grown plants require materials and devices that are devoid of any disease or infections. Minimal plant support elements are an essential component of a true aeroponic culture and apparatus. By reducing the amount of contact between the plant and the support structure, this design maximizes the airflow surrounding the plant. Ensuring the root systems are unrestricted is essential for long-term aeroponic cultivation because this will allow for unimpeded access to clean water, sufficient air exchange, and disease-free environments. In aeroponics, clean air is essential to the purification of the plant habitat. Air must be freely accessible for plants to grow naturally and for their

DOI: 10.4018/979-8-3693-2320-5.ch001

physiological development to be successful. So, the development of aeroponics, as well as its fundamental ideas and important techniques, are briefly discussed in this chapter.

INTRODUCTION

Aeroponics is an advanced method of growing plants without soil or any solid growing medium. Instead, plants are grown with their roots suspended in the air, and a nutrient-rich mist is periodically sprayed directly onto the roots. This method is part of a broader category of soilless cultivation techniques, which also includes hydroponics and aquaponics.

Aeroponics was first developed in the 1940s, but it gained significant attention in the 1980s when NASA began exploring its potential for growing food in space. The method's ability to maximize efficiency in water and nutrient use while reducing the need for heavy growing media made it an attractive option for space missions. Ending hunger and poverty while ensuring the sustainability of agriculture and food systems is one of the biggest issues of our day. Nonetheless, our top priority is to provide the future generation with wholesome, fresh food, especially with the world's expanding population (Alexandratrs and Bruinsma,2012). Global food production is increasing faster than population growth and per capita consumption. According to studies, the world's population is predicted to reach 10 billion by 2050—a 34% increase from the current number. The developing world will account for the majority of the population growth (Cohen, 2002; UN, 2010). The high population concentration has significant socioeconomic effects on food production, supply, and security, according to FAO (2009, 2011) reports, which call for more investigation. It will also emphasize the various issues, difficulties, and factors contributing to the rise in the number of people who are undernourished and hungry. By 2050, an extra 60% to 70% of the world's food supply will be required to sustain the growing and more urban population (Foote,2015).

Aeroponics represents a cutting-edge approach to agriculture that offers numerous benefits in terms of efficiency and sustainability. While it presents certain challenges, particularly in terms of cost and complexity, its potential for use in urban farming, space exploration, and other innovative agricultural practices continues to make it an area of active research and development. As the global population continues to grow, and the demand for sustainable food production increases, aeroponics is likely to play an increasingly important role. Ongoing research is focused on improving the efficiency, cost-effectiveness, and scalability of aeroponics systems, making them more accessible for widespread agricultural use. Aeroponics continues to be a key area of interest for space agencies as they plan for long-term space missions and potential colonization of other planets. The technology's ability to grow food with limited resources makes it ideal for space environments is aeroponics as shown in the following Figure 1.

Figure 1. Aeroponics plant growth

HISTORY OF AEROPHONICS

The history of aeroponics is a fascinating journey that spans several decades, with roots in agricultural innovation and scientific research. In the early 1940s, The basic concept of growing plants without soil was explored as part of the broader study of hydroponics. Researchers began experimenting with various methods to provide plants with the necessary nutrients without using soil. These early experiments laid the groundwork for what would later become aeroponics. During the year of 1966, The term "aeroponics" was first coined by Dr. Frank W. Went, a Dutch plant physiologist, in a publication titled The Experimental Control of Plant Growth.

Dr. Went's work focused on understanding the effects of environmental factors on plant growth, and his experiments with air and mist-based systems were among the first formal explorations of aeroponics. During the period (1970 – 1980), researchers began to refine the techniques of aeroponics. The technology started gaining attention as a method to study plant root systems in detail, as the roots were more accessible for observation and experimentation compared to soil-based systems as shown in Figure 2. At the 1980s NASA's interest in aeroponics significantly advanced the technology. As part of their Controlled Ecological Life Support System (CELSS) program, NASA explored aeroponics as a way to grow food in space. The space agency recognized the potential of aeroponics to maximize resource efficiency—critical for long-duration space missions where water and nutrients are limited.

In 1982, NASA scientist Richard Stoner developed the first commercial aeroponics system, which he called the "Genesis Growing System." This system was used to grow plants in a controlled environment and demonstrated the practicality of aeroponics for high-efficiency agriculture. Similarly, in 1990, NASA continued its research into aeroponics, focusing on optimizing the system for use in space. This research helped improve the technology, making it more reliable and efficient. The work conducted by NASA also highlighted the potential for aeroponics to be used in harsh or resource-scarce environments on Earth, such as deserts or urban areas. There is a drastic development in the 20th century, at the 2000s, Aeroponics began to see more widespread commercial adoption, particularly in the realm of vertical farming and urban agriculture. The ability to grow plants in controlled environments with minimal water and nutrient use made aeroponics an attractive option for sustainable farming practices. By 2010s-present, The rise of interest in sustainable and urban agriculture has further driven the development and adoption of aeroponics. Companies and startups have developed sophisticated aeroponic systems that are used in everything from small-scale home gardens to large commercial farms. This technology is now seen as a key component of vertical farming, where space is at a premium, and efficiency is crucial.

The history of aeroponics is marked by innovation and adaptation, from its early conceptual beginnings to its modern-day applications in sustainable agriculture and space exploration. The technology continues to evolve, driven by the need for more efficient and sustainable farming methods. As we look to the future, aeroponics holds promise as a solution to some of the most pressing challenges in agriculture and beyond.

Figure 2. Evolution of aeroponics

METHODS OF AEROPONICS

Aeroponics is a versatile and innovative method of growing plants in an air or mist environment without the use of soil or any other aggregate medium. There are several methods of aeroponics, each with its own approach to delivering nutrients and maintaining optimal conditions for plant growth. There are Six key methods as shown in the following Figure.3

Figure 3. Key methods of aeroponics

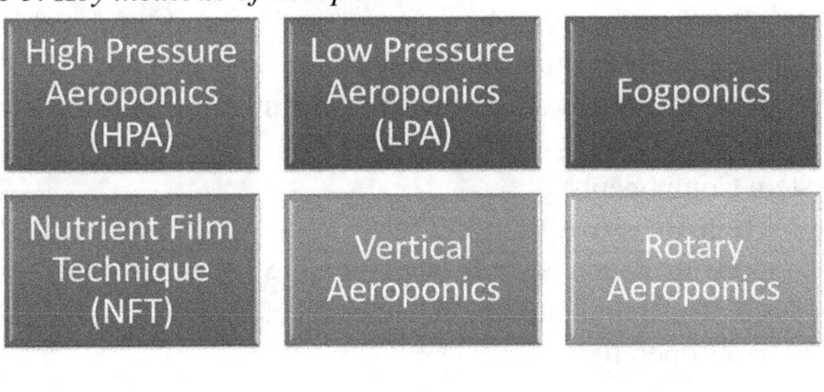

High Pressure Aeroponics (HPA)

High-pressure aeroponics (HPA) is an advanced method of aeroponics that involves delivering a fine mist of nutrient-rich water directly to the roots of plants at high pressure. This technique is known for its efficiency and effectiveness in promoting rapid plant growth and high yields, making it a popular choice for commercial growers and advanced hobbyists. The Key Features of High-Pressure Aeroponics are as follows

1. Mist Generation
   - **Pressure Requirements**: In HPA, the nutrient solution is atomized using a high-pressure pump, typically operating at 60-100 psi (pounds per square inch). This high pressure forces the nutrient solution through specialized mist nozzles that create a fine mist, with droplets usually around 50 microns or smaller.
   - **Droplet Size**: The small droplet size ensures that the roots are evenly coated with nutrients while also allowing for maximum oxygen absorption, which is crucial for healthy root development.
2. Nutrient Delivery
   - **Nutrient Solution**: The mist in an HPA system contains a carefully balanced mixture of water and nutrients tailored to the specific needs of the plants. The solution is delivered in short bursts, typically lasting a few seconds, followed by a pause to allow the roots to absorb the nutrients and oxygen.
   - **Frequency**: The misting cycle is typically set to occur every few minutes, depending on the specific needs of the plants and the environmental conditions. This intermittent misting helps prevent root desiccation while maintaining optimal moisture levels.
3. System Components
   - **High-Pressure Pump**: The core of an HPA system is a high-pressure pump capable of maintaining consistent pressure levels. This pump must be reliable and capable of handling continuous operation.

- **Mist Nozzles**: Specialized nozzles designed for high-pressure systems are essential for creating the fine mist. These nozzles must be durable and resistant to clogging, as even small blockages can disrupt the misting process.
- **Control System**: A sophisticated control system is often used to manage the timing and duration of misting cycles. This system may include timers, sensors, and automated controllers to ensure precise delivery of nutrients.

4. Advantages of HPA
    - **Efficient Water and Nutrient Use**: HPA systems use water and nutrients more efficiently than other growing methods, reducing waste and lowering operational costs.
    - **Rapid Growth**: The fine mist allows roots to absorb nutrients more quickly and efficiently, leading to faster plant growth and higher yields.
    - **Optimal Oxygenation**: The small droplet size and intermittent misting provide roots with ample oxygen, reducing the risk of root rot and promoting healthier plants.
    - **Disease Prevention**: With roots suspended in air and regularly misted, the risk of soil-borne diseases is minimized.
5. Challenges of HPA
    - **High Initial Cost**: The equipment required for HPA, including high-pressure pumps and specialized nozzles, can be expensive. This makes HPA more suitable for commercial operations or dedicated hobbyists.
    - **Maintenance**: The system requires regular maintenance to ensure that pumps and nozzles remain in good working order. Any malfunction can quickly lead to plant stress or failure.
    - **Technical Complexity**: HPA systems require precise control over misting cycles, nutrient concentrations, and environmental conditions. This can be challenging for beginners and requires a good understanding of plant physiology and system management.

Applications of High-Pressure Aeroponics

- **Commercial Farming**: Due to its efficiency and potential for high yields, HPA is widely used in commercial farming, particularly for high-value crops like herbs, leafy greens, and specialty vegetables.
- **Research**: HPA systems are also used in research settings, where precise control over growing conditions is required to study plant growth and development.
- **Urban and Vertical Farming**: HPA is increasingly popular in urban and vertical farming operations, where space and resource efficiency are critical.

High-pressure aeroponics (HPA) represents the cutting edge of soilless cultivation, offering significant advantages in terms of efficiency, growth rates, and yield. However, the complexity and cost of setting up and maintaining an HPA system mean that it is best suited for commercial operations or experienced growers looking to maximize their production in a controlled environment as shown in Figure 4.

Figure 4. High-pressure aeroponics

Inference

High-pressure aeroponics offers a more sophisticated and efficient growing environment, but its higher costs and technical complexity make it less adaptable for new users or small-scale growers. This method is better suited to experienced growers or commercial operations that can justify the investment with large-scale, high-yield production. However, the precise control it provides can yield faster growth and higher efficiency, making it a powerful tool for high-value crops or research purposes.

Low Pressure Aeroponics (LPA)

Low-pressure aeroponics (LPA) is a simpler and more accessible form of aeroponics compared to High-Pressure Aeroponics (HPA). It involves delivering a nutrient-rich mist to plant roots using lower-pressure systems, making it easier and more cost-effective to implement. The key features necessary for low-pressure aeroponics are given below.

Key Features of Low-Pressure Aeroponics (LPA)

1. Mist Generation
    - **Pressure Requirements:** LPA systems operate at lower pressures compared to HPA, typically between 10-30 psi (pounds per square inch). This results in a coarser mist or larger droplets compared to the fine mist produced by HPA systems.
    - **Droplet Size:** The droplets in an LPA system are larger (often over 100 microns) than those in HPA systems, which can affect nutrient absorption and root oxygenation.
2. Nutrient Delivery
    - **Nutrient Solution:** Like HPA, LPA uses a nutrient solution mixed with water, but the mixture is delivered through a less refined mist or spray. The solution is distributed intermittently, with misting cycles set to occur at regular intervals.
    - **Frequency:** The misting frequency can be adjusted based on the needs of the plants and the environmental conditions, but the cycles are generally longer compared to HPA systems.

3. System Components
    - **Low-Pressure Pump:** The system relies on a low-pressure pump to create the mist or spray. These pumps are typically less expensive and easier to maintain than high-pressure pumps.
    - **Spray Nozzles:** LPA systems use nozzles designed for lower pressure, which produce larger droplets. These nozzles are simpler and less prone to clogging than those used in high-pressure systems.
    - **Control System:** While LPA systems still require control over misting cycles, the setup is generally less complex. Basic timers or controllers are often sufficient for managing misting intervals.
4. Advantages of LPA
    - **Cost-Effective:** LPA systems are generally less expensive to set up and maintain than high-pressure systems. The components are more affordable, and the system is simpler to operate.
    - **Ease of Use:** LPA systems are easier to install and manage, making them suitable for hobbyists and small-scale growers.
    - **Reduced Complexity:** The lower pressure and larger droplets reduce the likelihood of nozzle clogging and system failures, simplifying maintenance.
5. Disadvantages of LPA
    - **Less Efficient Nutrient Absorption:** The coarser mist or larger droplets in LPA systems may not provide the same level of nutrient absorption and root oxygenation as HPA systems. This can lead to slower growth rates and lower yields.
    - **Risk of Overwatering:** Larger droplets can lead to waterlogging if misting cycles are not carefully managed, potentially causing root rot or other issues.
    - **Lower Precision:** The system offers less precise control over misting and nutrient delivery compared to high-pressure systems.

Applications of Low-Pressure Aeroponics

- **Small-Scale and Hobbyist Growing:** Due to its simplicity and lower cost, LPA is often used by hobbyists and small-scale growers who want to experiment with aeroponics without the high investment required for high-pressure systems.
- **Educational and Research Settings:** LPA systems can be used in educational settings to teach the principles of aeroponics and plant growth without requiring complex equipment.
- **Urban and Indoor Farming:** LPA can be used in urban and indoor farming operations where space is limited, and cost is a concern.

Low-Pressure Aeroponics (LPA) provides a more accessible and cost-effective alternative to High-Pressure Aeroponics (HPA). While it may not offer the same level of efficiency and precision as HPA, LPA is suitable for smaller-scale applications and beginners. It strikes a balance between ease of use and effectiveness, making it a practical choice for a variety of aeroponic growing scenarios as shown in the following Figure 5.

Figure 5. Low pressure aeroponics

Inference

This method is the most adaptable for beginners or small-scale operations. Its simplicity and low cost make it ideal for those new to aeroponics or those looking to experiment with it on a smaller budget. While it may not provide the most optimal mist for large-scale or highly sensitive crops, it is highly practical for hobbyists and entrepreneurs testing the waters of aeroponic farming. The setup can be scaled over time, making it versatile for a range of applications.

Fogponics

Fogponics is a variation of aeroponics that uses ultrasonic foggers to produce an extremely fine mist, or fog, which is then delivered to plant roots. This method creates a very fine nutrient mist with droplets typically ranging from 1 to 10 microns in size.

Key Features of Fogponics

1. Mist Generation
    - **Ultrasonic Foggers**: Fogponics systems use ultrasonic foggers to generate the fog. These devices use ultrasonic waves to break down the nutrient solution into very fine droplets. The foggers operate at a high frequency (typically around 1.7 million Hz) to create a dense fog.
    - **Droplet Size**: The fog produced is much finer than that of high-pressure or low-pressure aeroponics systems. The very small droplet size ensures that nutrients are delivered efficiently and that oxygen is readily available to the roots.
2. Nutrient Delivery
    - **Nutrient Solution**: The fog is composed of a nutrient-rich solution that is mixed with water. The fine particles in the fog allow for efficient nutrient uptake by the plant roots.
    - **Distribution**: The fog is typically dispersed throughout the root zone, enveloping the roots and providing a constant supply of nutrients and moisture.

3. System Components
   - **Ultrasonic Foggers**: These are the key components in fogponics. They must be high-quality and capable of consistently producing the desired fog density.
   - **Fog Chamber**: A chamber or enclosure where the fog is generated and where the plant roots are exposed to the mist. The chamber needs to be sealed to maintain optimal humidity and temperature levels.
   - **Control System**: Monitoring and controlling the fogging cycles and nutrient concentrations is essential. This may involve timers, sensors, and automated controllers to ensure consistent and accurate delivery.
4. Advantages of Fogponics
   - **Enhanced Nutrient Absorption**: The ultra-fine mist allows for superior nutrient uptake and better root oxygenation, which can lead to faster plant growth and higher yields.
   - **Optimal Oxygen Availability**: The fine droplets and the fog environment provide abundant oxygen to the roots, reducing the risk of root diseases and promoting healthy plant development.
   - **Water Efficiency**: Fogponics uses very little water compared to traditional growing methods, as the water is dispersed in the form of fog and can be recycled.
5. Disadvantages of Fogponics
   - **Equipment Cost and Maintenance**: Ultrasonic foggers and other components can be costly and require regular maintenance. Foggers can be prone to malfunction and may need frequent cleaning to prevent clogging.
   - **Technical Complexity**: The setup and operation of a fogponics system require precise control over fog density, nutrient concentration, and environmental conditions. This complexity can be challenging for beginners.
   - **Humidity Control**: Maintaining the right humidity levels within the fog chamber is crucial. Too much humidity can lead to issues like Mold or fungal growth, while too little can cause the roots to dry out.

Applications of Fogponics

- **Research and Development**: Fogponics is used in research settings to study plant physiology and root systems due to its ability to provide detailed control over the growing environment.
- **Commercial Farming**: Some high-tech commercial growers use fogponics for high-value crops where rapid growth and efficient nutrient delivery are critical.
- **Urban and Vertical Farming**: Fogponics is suitable for urban and vertical farming environments where space and water efficiency are important considerations.

Fogponics represents a sophisticated approach to soilless cultivation, offering benefits in terms of nutrient absorption and oxygenation. However, the complexity and cost of the equipment, along with the need for precise environmental control, make it a more advanced option compared to other aeroponic methods. Despite these challenges, fogponics holds significant potential for high-efficiency growing systems and specialized applications as shown in following Figure 6.

Figure 6. Fogponics

Inference

Fogponics is an advanced form of aeroponics, offering a fine mist that can provide superior nutrient uptake and growth potential. However, the complexity and cost of maintaining ultrasonic foggers, as well as the more specialized conditions required, make it less adaptable for the average grower. It is better suited for specific use cases like growing delicate or high-value plants in controlled environments, but may not be practical for broader commercial adoption without considerable expertise.

Nutrient Film Technique (NFT)

The Nutrient Film Technique (NFT) is a type of hydroponic growing method where a thin film of nutrient-rich water continuously flows over the roots of plants. This technique is widely used in soilless cultivation due to its efficiency and effectiveness.

Key Features of the Nutrient Film Technique (NFT)

1. System Design
   - **Growing Channels**: NFT systems typically use sloped, open channels or troughs where the plant roots are exposed to a thin film of nutrient solution. These channels are often made from materials like PVC or other non-porous substances.
   - **Nutrient Film**: A shallow film of nutrient solution continuously flows over the roots, providing them with the essential nutrients and oxygen. The film is usually just a few millimeters deep.
2. Nutrient Delivery
   - **Pumping System**: A pump is used to circulate the nutrient solution from a reservoir to the top of the channels. The solution then flows down the channels, returning to the reservoir for recirculation.
   - **Flow Rate**: The flow rate of the nutrient solution is crucial. It must be high enough to ensure that the film of solution remains

consistent but not so high that it causes waterlogging or nutrient imbalances.

3. Advantages of NFT
   - **Efficient Nutrient Use**: NFT systems use less water and nutrients compared to traditional soil-based growing methods because the solution is recirculated, reducing waste.
   - **Rapid Growth**: The constant exposure to nutrients and oxygen can lead to faster plant growth and higher yields.
   - **Reduced Disease Risk**: The lack of soil reduces the risk of soil-borne diseases and pests.
   - **Space Efficiency**: NFT systems can be set up vertically, maximizing space usage and making them suitable for urban and indoor farming.
4. Disadvantages of NFT
   - **System Vulnerability**: NFT systems are highly dependent on the continuous flow of nutrient solution. Any disruption in the flow, such as pump failure or clogging, can quickly lead to plant stress or death.
   - **Monitoring and Maintenance**: Regular monitoring of nutrient levels, pH, and flow rate is necessary to maintain optimal growing conditions. Any imbalances can affect plant health and growth.
   - **Limited Plant Types**: NFT is best suited for certain types of plants, such as leafy greens and herbs. Plants with larger root systems or those requiring more substantial support may not be ideal for NFT systems.
5. System Variations
   - **Horizontal NFT**: The most common design, where channels are positioned horizontally, allowing the nutrient solution to flow evenly over the roots.
   - **Vertical NFT**: Channels are positioned vertically, which is useful for space-saving in urban or indoor environments. The nutrient solution flows down from the top of the channels, cascading over the roots.
   - **Deep Flow Technique (DFT)**: A variation of NFT where the nutrient film is deeper, providing more nutrient solution to the

roots. This can help with larger plants or those with more extensive root systems.

Applications of NFT

- **Commercial Agriculture**: NFT is widely used in commercial hydroponic farms for growing crops like lettuce, spinach, and herbs. Its efficiency and space-saving design make it a popular choice for large-scale operations.
- **Urban Farming**: The compact and vertical design of NFT systems makes them ideal for urban farming setups, where space is limited.
- **Research**: NFT systems are used in research settings to study plant growth and nutrient uptake due to their precise control over growing conditions.

The Nutrient Film Technique (NFT) offers a highly efficient and space-saving method for growing plants hydroponically. While it presents some challenges, particularly in terms of system reliability and plant suitability, it remains a popular choice for both commercial and experimental hydroponic systems. Proper management and maintenance are key to ensuring the success of NFT systems and achieving optimal plant growth as shown in Figure 7.

Figure 7. Nutrient film technique (NFT)

Inference

The NFT system is highly efficient in terms of water and nutrient usage since it recirculates the nutrient solution, minimizing waste. This makes it cost-effective for large-scale operations, particularly in commercial environments. The system is also scalable; it can be expanded for large commercial farms or downsized for small hobbyists. However, NFT is more suited for plants with smaller root systems, as large root systems can clog the channels and disrupt the nutrient flow.

Vertical Aeroponics

Vertical Aeroponics is an advanced method of aeroponics where plants are grown in a vertical arrangement, allowing for maximized space utilization. This system involves stacking multiple layers of plants vertically, with nutrient mist delivered to their roots. This technique is particularly useful in urban environments or areas with limited horizontal space.

Key Features of the Nutrient Film Technique (NFT)

1. System Design
    - **Growing Channels**: NFT systems typically use sloped, open channels or troughs where the plant roots are exposed to a thin film of nutrient solution. These channels are often made from materials like PVC or other non-porous substances.
    - **Nutrient Film**: A shallow film of nutrient solution continuously flows over the roots, providing them with the essential nutrients and oxygen. The film is usually just a few millimeters deep.
2. Nutrient Delivery
    - **Pumping System**: A pump is used to circulate the nutrient solution from a reservoir to the top of the channels. The solution then flows down the channels, returning to the reservoir for recirculation.
    - **Flow Rate**: The flow rate of the nutrient solution is crucial. It must be high enough to ensure that the film of solution remains

consistent but not so high that it causes waterlogging or nutrient imbalances.

3. Advantages of NFT
    - **Efficient Nutrient Use**: NFT systems use less water and nutrients compared to traditional soil-based growing methods because the solution is recirculated, reducing waste.
    - **Rapid Growth**: The constant exposure to nutrients and oxygen can lead to faster plant growth and higher yields.
    - **Reduced Disease Risk**: The lack of soil reduces the risk of soil-borne diseases and pests.
    - **Space Efficiency**: NFT systems can be set up vertically, maximizing space usage and making them suitable for urban and indoor farming.
4. Disadvantages of NFT
    - **System Vulnerability**: NFT systems are highly dependent on the continuous flow of nutrient solution. Any disruption in the flow, such as pump failure or clogging, can quickly lead to plant stress or death.
    - **Monitoring and Maintenance**: Regular monitoring of nutrient levels, pH, and flow rate is necessary to maintain optimal growing conditions. Any imbalances can affect plant health and growth.
    - **Limited Plant Types**: NFT is best suited for certain types of plants, such as leafy greens and herbs. Plants with larger root systems or those requiring more substantial support may not be ideal for NFT systems.
5. System Variations
    - **Horizontal NFT**: The most common design, where channels are positioned horizontally, allowing the nutrient solution to flow evenly over the roots.
    - **Vertical NFT**: Channels are positioned vertically, which is useful for space-saving in urban or indoor environments. The nutrient solution flows down from the top of the channels, cascading over the roots.
    - **Deep Flow Technique (DFT)**: A variation of NFT where the nutrient film is deeper, providing more nutrient solution to the

roots. This can help with larger plants or those with more extensive root systems.

Applications of NFT

- **Commercial Agriculture**: NFT is widely used in commercial hydroponic farms for growing crops like lettuce, spinach, and herbs. Its efficiency and space-saving design make it a popular choice for large-scale operations.
- **Urban Farming**: The compact and vertical design of NFT systems makes them ideal for urban farming setups, where space is limited.
- **Research**: NFT systems are used in research settings to study plant growth and nutrient uptake due to their precise control over growing conditions.

The Nutrient Film Technique (NFT) offers a highly efficient and space-saving method for growing plants hydroponically. While it presents some challenges, particularly in terms of system reliability and plant suitability, it remains a popular choice for both commercial and experimental hydroponic systems. Proper management and maintenance are key to ensuring the success of NFT systems and achieving optimal plant growth as shown in Figure 8.

Figure 8. Vertical aeroponics

Inference

One of the major advantages of vertical aeroponics is its optimal use of space. By growing plants vertically, this method maximizes the use of vertical real estate, making it highly efficient for urban agriculture or areas with limited growing space. It allows for high-density plant production in small areas, which is particularly useful for urban farming, indoor growing, and greenhouse environments. For commercial operations, this can significantly increase the number of plants grown per square foot, improving yield without requiring a large land area.

Rotary Aeroponics

Rotary Aeroponics is an innovative approach to aeroponics where plants are grown in a rotating drum or wheel. This method combines the principles of aeroponics with rotational motion to enhance nutrient delivery and plant growth.

Key Features of Rotary Aeroponics

1. System Design
    - **Rotating Drum or Wheel**: In a rotary aeroponics system, plants are mounted on a rotating drum or wheel. The drum or wheel continuously rotates, exposing the plant roots to a nutrient mist as it turns.
    - **Rotation Mechanism**: The rotation is typically powered by a motor or belt system, which ensures a consistent and smooth rotation of the drum or wheel. The rotation can be set at various speeds depending on the needs of the plants.
2. Nutrient Delivery
    - **Misting System**: Nutrient-rich mist is delivered to the roots through nozzles or sprayers positioned around the rotating drum. As the drum rotates, the mist is evenly distributed over the roots.
    - **Continuous Exposure**: The rotation ensures that all parts of the root system are exposed to the nutrient mist, providing uniform nutrient and oxygen delivery.
3. Advantages of Rotary Aeroponics
    - **Uniform Nutrient Distribution**: The rotation ensures that the nutrient mist reaches all parts of the root system, promoting uniform growth and nutrient absorption.
    - **Enhanced Root Oxygenation**: The rotating motion helps to increase air circulation around the roots, which improves oxygen availability and reduces the risk of root diseases.
    - **Space Efficiency**: Rotary systems can be designed to maximize vertical space, making them suitable for environments where horizontal space is limited.
    - **Increased Growth Rates**: The combination of aeroponics and rotational motion can lead to faster plant growth and higher yields due to efficient nutrient and oxygen delivery.
4. Disadvantages of Rotary Aeroponics
    - **Complexity and Cost**: The design and construction of rotary aeroponics systems can be complex and costly. The rotating

mechanism and misting system require careful engineering and maintenance.

- **Mechanical Wear**: The rotating components, such as bearings and motors, can be subject to wear and may require regular maintenance or replacement.
- **System Maintenance**: Ensuring that the misting system and rotating components function properly requires regular monitoring and maintenance. Any malfunction in the rotation or misting can impact plant health.

5. Applications of Rotary Aeroponics
   - **Commercial Production**: Rotary aeroponics is used in commercial hydroponic farms to increase efficiency and productivity. It is particularly useful for high-value crops where rapid growth and high yields are desired.
   - **Research**: This method is used in research settings to study plant growth and nutrient uptake under controlled conditions. The ability to precisely control nutrient delivery and oxygenation makes it valuable for scientific studies.
   - **Urban and Indoor Farming**: The space-efficient design of rotary systems makes them suitable for urban and indoor farming environments, where maximizing space and resource use is critical.
6. System Variations
   - **Horizontal Rotating Drums**: Plants are grown on horizontal drums that rotate around a central axis. The nutrient mist is delivered through nozzles positioned around the drum.
   - **Vertical Rotating Drums**: Plants are grown on vertical drums or wheels, which rotate around a vertical axis. This design can be used to save floor space and increase plant density.

Rotary Aeroponics offers a unique approach to soilless cultivation by combining rotational motion with aeroponic nutrient delivery as shown in Figure 9. While it presents some challenges in terms of complexity and cost, it provides significant advantages in terms of uniform nutrient distribution, enhanced oxygenation, and space efficiency. With careful design and maintenance, rotary aeroponics can be an effective solution for high-efficiency growing systems in both commercial and research applications.

Figure 9. Rotary aeroponics

Inference

Rotary aeroponics is highly efficient in terms of space utilization. The rotating mechanism allows plants to be grown in a compact vertical structure, which is ideal for indoor environments and urban farming where land is limited. By using a cylindrical design, plants can be arranged in a circular fashion around a central light source, ensuring even light distribution and maximizing the number of plants grown per square foot. This makes rotary aeroponics especially useful for high-density farming in constrained spaces like greenhouses or indoor vertical farms.

CHALLENGES ASSOCIATED WITH IMPLEMENTING AEROPONIC SYSTEMS

Challenges associated with implementing aeroponic systems include:

- **High Initial Investment Costs:** The setup for aeroponic systems can be expensive, which may deter potential adopters.

- **Technical Complexity:** The engineering and design of aeroponic systems require specialized knowledge, making them more complex to operate compared to traditional methods.
- **Nutrient Delivery Complexities:** Ensuring that plants receive the correct nutrients in an aeroponic system can be challenging.
- **Equipment Issues:** The reliance on technology means that equipment failures can significantly impact plant health and growth.
- **Higher Quality Water Needed:** Aeroponic systems require a higher quality of water, which can be a limiting factor in some areas.
- **Difficulties in Organic Cultivation:** Growing plants fully organically in an aeroponic system can be more challenging compared to soil-based systems.
- **Rapid Deterioration in Plant Health:** When problems arise, such as nutrient imbalances or system failures, plant health can decline rapidly

REFERENCES

Alexandratrs, N., & Bruinsma, J. (2012). World agriculture towards 2030/2030. The 2012 revision. Global perspective Studies. FAO, ESA Working Paper No. 12-03. https://www.fao.org/agrifood-economics/en/.

Brockerhoff, M., & Nations, U.United Nation. (2010). World urbanization prospects: The 2009 revision. *Population and Development Review*, *24*(4), 883. DOI: 10.2307/2808041

Chang, D. C., Park, C. S., Kim, S. Y., Kim, S. J., & Lee, Y. B. (2008). Physiological growth responses by nutrient interruption in aeroponically grown potatoes. *American Journal of Potato Research*, *85*(5), 315–323. DOI: 10.1007/s12230-008-9024-4

Chang, D. C., Park, C. S., Kim, S. Y., & Lee, Y. B. (2012). Growth and tuberization of hydroponically grown potatoes. *Potato Research*, *55*(1), 69–81. DOI: 10.1007/s11540-012-9208-7

Cohen, J. (2002). World population in 2050: assessing the projections. In Little, J. S., & Triest, R. K. (Eds.), *Seismic shifts: the economic impact of demographic change, Federal Reserve Bank of Boston conference. Series No. 46:83–113*.

Food and Agricultural Organizations of the United Nations. (2011) Food, agriculture and cities. Challenges of food and nutrition security, agriculture and ecosystem management in an urbanizing world.

FAO Food and Agriculture Organization. (2009) Report, global agriculture towards 2050. https://www.fao.org/fileadmin/templates/wsfs/docs/Issues_papers/HLEF2050_Global_Agriculture.pdf

Foote, W. (2015). *To feed the world in 2050, we need to view small-scale farming as a business*. Skoll World Forum.

Kamies, R., Rafudeen, M. S., & Farrant, J. (2010). The use of aeroponics to investigate antioxidant activity in the roots of Xerophytaviscosa. *Plant Growth Regulation*, *62*(3), 203–211. Advance online publication. DOI: 10.1007/s10725-010-9498-1

Mbiyu, M. W., Muthoni, J., Kabira, J., Elmar, G., Muchira, C., Pwaipwai, P., Ngaruiya, J., Otieno, S., & Onditi, J. (2012). Use of aeroponics technique for potato (Solanum tuberosum) minitubers production in Kenya. *Journal of Horticulture and Forestry*, *4*, 172–177.

Osvald, J., Petrovic, N., & Demsar, J. (2001). Sugar and organic acid content of tomato fruits (lycopersicon lycopersicum mill.) grown on aeroponics at different plant density. *Acta Alimentaria*, *30*(1), 53–61. DOI: 10.1556/AAlim.30.2001.1.6

Song, Y., Ma, J., Zhang, X., & Feng, Y. (2012). Design of wireless sensor network-based greenhouse environment monitoring and automatic control system. *J of Netw.*, *7*(5), 838–844. DOI: 10.4304/jnw.7.5.838-844

Tik, L. B., Khuan, C. T., & Palaniappan, S. (2009). Monitoring of an aeroponic greenhouse with a sensor network, IJCSNS. *Inter J of Compu Sci and Net Sec.*, *9*(3), 240–246.

Chapter 2
Outline of Aeroponic Technique

Mohana Priya Rajendran
SNS College of Technology, India

R. Muthuminal
https://orcid.org/0000-0001-9931-2204
SNS College of Technology, India

R. Vigneshkumar
SNS College of Technology, India

A. Sanakkiyan
SNS College of Technology, India

ABSTRACT

Finding sustainable solutions is becoming more and more important for the agricultural sector as environmental pollution and climate change worsens. Through interventions of technologies such as ICT, sensor and actuator technology made smart farming easier. But there have been difficulties along the way with smart farming, especially with indoor hydroponic systems. High costs and the inability to compete with the pricing of traditionally grown commodities have forced many businesses to file for bankruptcy. So, to overcome the issues the researchers came up with

DOI: 10.4018/979-8-3693-2320-5.ch002

a renowned idea called Aeroponics. This chapter elaborated the History, Classification, Working, Advantage, Disadvantage along with difference between hydroponic and aeroponic.

INTRODUCTION

Finding sustainable solutions is becoming more and more important for the agricultural sector as environmental pollution and climate change worsens. Through the integration of information and communication technology (ICT) into agriculture, smart farming has become a viable method for optimizing crop growth conditions remotely or autonomously, overcoming spatial and temporal constraints (Knierim, A et.al., 2019, 7. ElBeheiry, N.; Balog, R.S.,2023). Rapid developments in sensor and actuator technology in the early 2000s made smart farming more widely available and economically viable, which fuelled its commercial growth.

But there have been difficulties along the way for smart farming, especially with indoor hydroponic systems. High costs and the inability to compete with the pricing of traditionally grown commodities have forced many businesses to file for bankruptcy (Campbell, C, 2023,Redmond,K. 2023, 10. Streed, A. et.al, 2019). This financial barrier has increased doubts about the sustainability of smart agricultural technology, as have the environmental problems associated with the disposal of waste solutions (Karunathilake, E.M.B.M. et.al,2023, Yépez-Ponce, D.F et.al, 2023, Nikolov, N.V. et.al, 2023, Ashraf, M.A. et.al, 2014). However, the demand for novel agricultural techniques is urgent due to the unabated advancement of climate change, environmental degradation, and groundwater depletion (Climate-Smart Agriculture,2024, Ampim, P.A. et.al, 2022, Lieder, S. et.al, 2021, Roy, T et.al,2020, Gnauer, C. et.al,2020, Wittmann, S. et.al,2020, Benke, K.et.al,2017).

Figure 1. Classification of plant cultivation techniques (Lakhiar et.al, 2020)

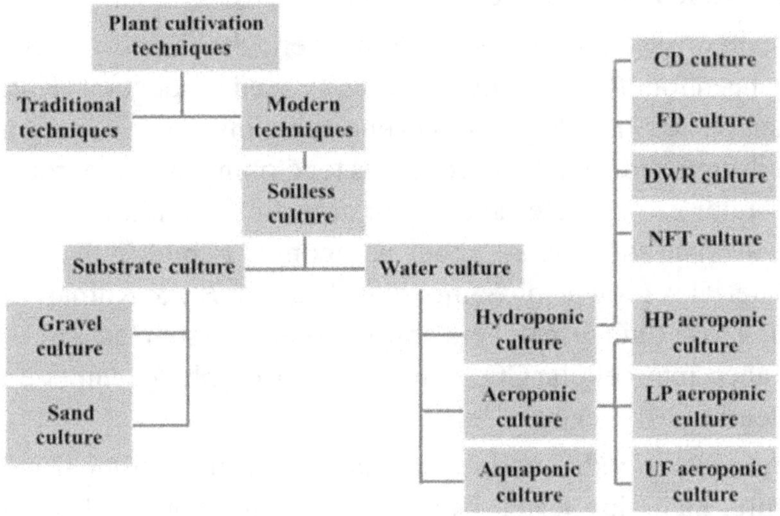

Note: CD: Continuous drip; FD: Flood and drain; DWR: Deep water raft culture; NFT: Nutrient film technique; HP: High-pressure; LP: low-pressure; UF: Ultrasonic fogger.

Aeroponics is a soil-free method of growing plants by suspending the plant roots in air and misting them with a nutrient-rich water solution. Aeroponic systems typically employ pumps, timers, and spray nozzles to provide a highly oxygenated mist of water and nutrients to plant roots. The mist is given at predetermined intervals, allowing the roots to absorb nutrients and moisture (Kumari, R., & Kumar, R. 2019).

Plants grown in aeroponic systems have also been found to absorb more minerals and vitamins, making them healthier and maybe more nutritious (Kumari, R., & Kumar, R. 2019). The suspended aeroponic plants receive 100% of the available oxygen and carbon dioxide to the roots, stems, and leaves, which accelerates biomass growth and shortens rooting times (Kumari, R., & Kumar, R. 2019). The increased biomass yield of aerial components from the aeroponic treatment suggested that this production technology should not only be used for root crops, but also for other types of crops. Furthermore, aeroponics allows for higher planting densities because plant-to-plant competition for nutrients and water is virtually eliminated. Any species of plants can be grown in a true aeroponic system because the micro-environment of an aeroponic can be finely controlled (Kumari, R., & Kumar, R. 2019).

Aeroponic seed production systems were developed in response to growing need for efficient and high-quality seed production (Kumari, R., & Kumar, R. 2019). Aeroponic bio-pharming involves growing pharmaceutical drugs inside plants. Aeroponics for cloning increases root growth, survival rate, growth rate, and maturation time. Studies have revealed that the average tuber yield under aeroponics is higher than when the same material is grown conventionally (Kumari, R., & Kumar, R. 2019). According to certain researchers, the aeroponics system boosted stomatal conductance, intercellular CO_2 concentration, net photosynthetic rate, and photochemical efficiency of leaves.

Aeroponics provides various advantages over traditional soil-based agriculture, including faster growth rates, larger crop yields, and more effective resource usage. The soil-free atmosphere also minimizes the risk of soil-borne illnesses and pests, making agriculture more sustainable and environmentally beneficial. However, its main drawbacks are the expensive initial setup costs, maintenance requirements, and a high level of technical competence required (Kumari, R., & Kumar, R. 2019).

Aeroponics is a relatively new technology that has already been used to effectively cultivate a variety of crops, including lettuce, tomatoes, and strawberries. One of the primary benefits of aeroponics technology is the ability to grow plants all year since environmental parameters like as humidity, temperature, airflow, and light intensity can be artificially adjusted using specific tools and systems (Kumari, R., & Kumar, R. 2019).

Figure 2. Aeroponic system (5)

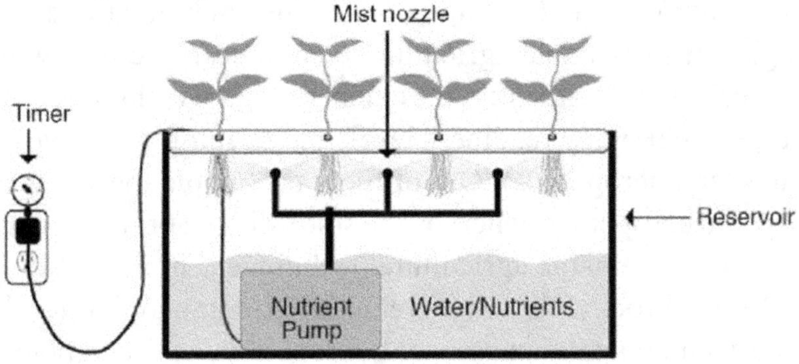

HISTORY

Aeroponic research began in the 1920s and has progressively advanced as a soilless growing technology. In the 1940s, technology was primarily used for research rather than commercial agricultural production. In 1942, W. Carter pioneered air culture cultivation by delivering nutrients to plant roots through water vapor, allowing for easier root investigation. In 1944, L.J. Klotz discovered vapor misted citrus plants while researching citrus and avocado root infections (Kumari, R., & Kumar, R. 2019). In 1952, G.F. Trowel used spray culture to grow apple trees. GTi's Genesis Rooting System, often known as the Genesis Machine, was the first commercial aeroponics arrangement introduced in 1983 (Kumari, R., & Kumar, R., 2019). The device was controlled by a microchip and connected to a power outlet and water faucet. NASA conducted tests in space and on Earth in the 1990s to grow biomass with minimal soil and water, resulting in high productivity (Kumari, R., & Kumar, R. 2019). NASA research found that aeroponically produced plants have 80% more dry weight biomass (essential minerals) than hydroponically grown plants. Aeroponic techniques have been given special attention from NASA since a mist is easier to handle than a liquid in a zero-gravity environment (Kumari, R., & Kumar, R. 2019). Some literature review studies have analyzed the uses of different technologies on the whole panorama of sustainable agriculture (Juan Garzón et.al,2023). For example, the study by Abbasi et al. (Juan Garzón et.al,2023) investigated the emerging trends of digital technologies in sustainable agriculture techniques. The review included 148 empirical studies to identify the extent of digital technology adoption, service type, technology readiness level, and farming type (hydroponics, aquaponics, and aeroponics). Out of the total sample, only three studies were related to aeroponics, whereby the study did not draw any conclusion on this type of sustainable agricultural technique. Similarly, the study by Fussy and Papenbrock (Juan Garzón et.al,2023) compared soil and soilless cultivation techniques (aquaponics, hydroponics, and aeroponics). The study concludes that implementing technological developments related to monitoring and automation gives the advantage to soilless techniques, making them much more profitable, well-organized, effective, secure, and nature-friendly. Finally, the study by Basso and Antle investigated the affordances of digital technologies such as sensors, robotics, AI, and data analytics as key developments to secure sustainable food production. The study argues that while complex climate changes put food production at risk, digital technologies are key to controlling such changes and achieving truly sustainable agriculture. However, as well as with the other studies, this one also did not specify any conclusions about the impact of technology on aeroponics (Juan Garzón et.al,2023).

Figure 3. Image of aeroponics (greentechendeavors.com)

Technologies Involved With Aeroponics

Aeroponics is a success because of technology, which makes it possible to precisely control all of which lead to increased food production. Additionally, compared to conventional farming methods, aeroponics may produce high-quality, nutrient-rich crops with a less environmental footprint by utilizing technology. (Juan Garzón et.al,2023)

In aeroponics, one of the main uses of technology is the monitoring of environmental variables through the use of sensors. In order to improve plant growth, farmers may make data-driven decisions about modifying environmental parameters, such as temperature, humidity, and nutrient levels, thanks to sensors, which offer real-time monitoring of these factors. This can involve monitoring the health of the plants, regulating humidity and temperature, and altering the amount of nutrients. Additionally, sensors can be combined with automated control systems to modify the environment in response to changing circumstances. By doing this, the possibility of human error is decreased and it is guaranteed that growing plants will always have optimal environmental conditions. Furthermore, sensors have the ability to identify problems before they are apparent to the human eye, such as nutrient deficits, pests, and illnesses. This early detection can help prevent crop loss and ensure that plants are healthy and productive (Juan Garzón et.al,2023).

Industry 4.0 is another technology option that has grown in importance in aeroponics. It alludes to the creation of intelligent manufacturing processes through the integration of cutting-edge technologies like the Internet of Things (IoT), artificial intelligence (AI), and data analytics. This is known as the fourth industrial revolution. Industry 4.0 technologies are becoming more and more significant in aeroponics since they provide a number of advantages for maximizing plant growth and enhancing efficiency. IoT can be used, for instance, to automate a variety of operations, including pest control, environmental control, and fertilizer supply. This guarantees that plants always have the best growing circumstances and lessens the need for human intervention. However, by examining plant photos, artificial intelligence (AI) may discover indications of plant illnesses that may go undetected by the human eye. This can assist farmers in acting quickly to stop the spread of illnesses and keep their crops healthy. Analyzing large volumes of data on plant development, weather patterns, and nutrient levels can also be done with data analytics, allowing for precise prediction of the best circumstances for plant growth. By doing this, farmers may increase yields and optimize their aeroponic systems (Juan Garzón et.al,2023).

Aeroponic systems also require dispenser-related technology because these devices provide plants with water and nutrients. This comprises water pumps, centrifugal pumps, atomizers, sprayers, nebulizers, and ultrasonic dispersion devices that are automatically controlled based on the demands identified by the sensors. Plants may be precisely dosed with nutrients and water thanks to dispenser technology. This guarantees that the plants get the right amount of water and nutrients, all of which are essential to their growth and development. Dispenser technology can also be programmed to automatically provide water and nutrients based on real-time sensor data or at predetermined intervals. This guarantees that the plants receive constant care and lessens the need for manual effort. To receive optimal nutrition to each plant, one can customize the dispenser for nutrient and water supply based on specific need(Juan Garzón et.al,2023).

Renewable energy-related technologies are another set of innovations that contribute to aeroponics' sustainability. Aeroponic systems can be powered by alternative energy sources, such solar or wind power, which has several advantages for lowering energy consumption and enhancing sustainability. This lessens the need for grid power, which can be costly and harmful to the environment. Growers can operate and maintain their aeroponic systems at a lower cost when they use renewable energy, as it lowers the energy expenses involved. For aeroponic producers, alternative energy sources are a more environmentally responsible option because they are usually cleaner and more sustainable than traditional ones(Juan Garzón et.al,2023).

Classification of Aeroponics

Figure 4. Classification of aeroponics

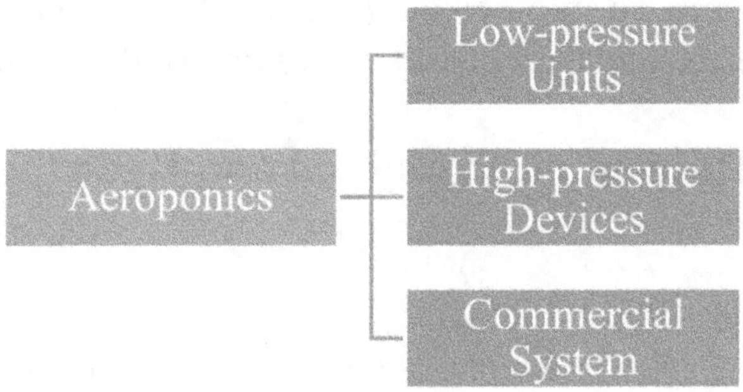

Low-Pressure Units

Low-pressure aeroponic gardens often hold plant roots above a nutrient solution reservoir or a canal that connects to it. A low-pressure pump delivers the nutritional solution through jets or ultrasonic transducers, which drip or drain back into the reservoir. When plants reach maturity, they develop dry root systems, preventing adequate nutrition uptake. Because of the expense, these machines lack functions for purifying the nutrient solution, removing debris, and eliminating undesirable pathogens. These units are often appropriate for benchtop growing. It is also used to demonstrate the fundamentals of aeroponics (Kumari, R., & Kumar, R. 2019).

High-Pressure Devices

High-pressure pumps generate mist in aeroponic systems. And it is commonly employed in the growth of high-value crops. This method involves technology for purifying air and water, sterilizing nutrients, using low-mass polymers, and delivering nutrients under pressure (Kumari, R., & Kumar, R. 2019).

Commercial System

The commercial system combines high-pressure device hardware with biological systems. The biological systems matrix includes an upgrade for plant life extension and crop maturation (Kumari, R., & Kumar, R. 2019).

Difference Between Aeroponics and Hydroponics

Table 1. Difference between hydroponic and aeroponic agriculture

Aspect	Hydroponic	Aeroponic
Growing Medium	Utilizes inert materials (like perlite, rockwool, or coconut coir) as support for plant roots within a nutrient-rich water solution.	Does not use a growing medium; plant roots are suspended in air and misted with a nutrient-rich solution.
Method of Nutrient Delivery	Nutrients delivered through the water solution that constantly circulates around the roots.	Nutrients are delivered directly to the roots via a fine mist or spray.
Water Usage	Generally uses more water compared to aeroponics due to the continuous flow or recirculation of water.	Highly water-efficient, using significantly less water compared to hydroponics due to the misting system and recycling capabilities.
Oxygen Levels for Roots	Good oxygen levels but not as high as aeroponics because roots are submerged in water.	Provides the highest oxygen levels to the roots as they are exposed to air, leading to faster growth rates.
Plant Growth	Growth rates are typically slower than aeroponics due to the indirect nutrient delivery system.	Offers faster growth rates and increased yields owing to direct nutrient delivery and optimal oxygenation of roots.
Susceptibility to Disease	More prone to diseases spread through the water system or growing medium.	Less prone to diseases due to the absence of a growing medium and exposure to open air.
Complexity of System	Relatively simpler system compared to aeroponics, easier for beginners to set up and maintain.	More complex system requiring precise misting, potentially challenging for novices without technical expertise.

continued on following page

Table 1. Continued

Aspect	Hydroponic	Aeroponic
Initial Investment	Initial setup costs are moderate and relatively lower compared to aeroponics.	Higher initial investment due to specialized misting systems and equipment.
Space Efficiency	Less space-efficient compared to aeroponics as plants require a growing medium.	Highly space-efficient, allowing vertical or stacked growing systems, maximizing space usage.

Source: Shashank et al. (2024)

Working

Aeroponic systems reduce agricultural labor due to their continuous operation in limited spaces. Aeroponics allows for the cultivation of vegetables without the need for a substratum or soil, as opposed to hydroponics. Instead, nutrients are delivered to the plants through containers. Aeroponic growing involves suspending plants in a closed or semi-closed environment and spraying their roots and lower stems with a nutrient-rich water solution (Kumari, R., & Kumar, R. 2019).

Aeroponic cultivation requires a monitoring and control system to ensure optimal water and fertilizer distribution. A nutrient solution storage tank requires a distribution system consisting of pipes, spray nozzles, a pump, and a timer to distribute spray(Kumari, R., & Kumar, R. 2019). It employs a small internal micro jet spray to spray the roots with a fine, high-pressure mist containing nutrient-rich solutions from the nutrient reservoir in the rooting chamber. There is a programmed cyclic timer that turns on the high-pressure aeroponic pump. Nutrients are mixed with water in a reservoir basin, which is then filtered and pumped into a pressurized holding tank that is intermittently misted onto the root system(Kumari, R., & Kumar, R. 2019).

Developed root hairs aid in the absorption of nutrients from the fluids. The plant's root system allows for easier nutrition delivery. The tiny size of the spray particles reduces nutritional solution waste significantly. And with a enough oxygen supply, root rot is absolutely avoided.

Misting is typically done every five minutes around the hung roots. The system would only turn on for a few seconds every 2-3 minutes. Interrupting the misting cycles can cause the roots to dry up quickly because to their exposure to air. A timer controls the nutrient pump similarly to other types of hydroponic systems, with the exception that the aeroponics system requires a short cycle timer that runs the pump for a few seconds every couple of minutes. However, the chamber must be made of lightless materials from all over, so that the roots are in darkness. This is also effective for inhibiting algae growth, which impedes plant growth and pollutes the system (Kumari, R., & Kumar, R. 2019).

The droplet size of a nutrient mist is an important factor in aeroponics. Oversized droplets may limit oxygen delivery. Undersized droplets can promote root hair growth and hinder lateral root growth, thus impacting aeroponic system performance. The water droplets must be large enough to transport nutrients to the roots in adequate quantities, but small enough not to precipitate out of the root mass. Unused solution drips down to the unit's base, where it is strained, filtered, and pumped back into the reservoir. Aeroponics systems enable easy monitoring of nutrients and pH. Aeroponics uses minimum contact between the support structure and the plant, allowing for unrestricted plant growth (Kumari, R., & Kumar, R. 2019).

Different Components of Aeroponics

1. Nutrients used in aeroponics: Mainly N-NH4 (0.54 g/L), N-NO3 (0.35 g/L), P(0.40 g/L), K (0.35 g/L), Ca (0.17 g/L), Mg(0.08 g/L), Na(0.04 g/L), Fe (0.09 g/L), Zn (0.03 g/L) and B(0.03g/L) are commercially being used in most of the crops (Kumari, R., & Kumar, R. 2019).
2. Water used in aeroponics: Water to be used in aeroponics should have a low EC, not exceeding one mS/cm. Water pH is also a useful indicator. Water sources with a pH of over eight are questionable for aeroponics. It is useful to have a water chemical analysis; even if EC and pH measures fall into acceptable levels. The other problem we may have to face is water biological contamination. Water from deep wells is usually not contaminated. Water from superficial wells, especially near urban areas, is likely to be contaminated with coli form bacteria,

including Pecto bacterium. Water from suspicious sources should have a microbiological analysis. Special filters can minimize this risk. If available, water should be Filtered before going into the nutrient tank. Boiling is also another alternative if no other is available (Kumari, R., & Kumar, R. 2019).

3. The plant material used in aeroponics: Optimum plant material should be used for aeroponics. In vitro plants are preferred because of sanitary reasons. However, they need to be handled with proper care by experienced technicians. These Plants should be the appropriate age and size and should go through a thorough acclimatization period before going into the greenhouse. Other plant materials, such as rooted cuttings and tuber sprouts, should be clean and disease free. The presence of any kind of symptom should be sufficient reason to discard the whole batch of plants. This should be noticeable when transplanting into the boxes. The underground part of the tissue coming from the sand trays should be completely clean and sand free. Before placing into aeroponics, plants should be managed in a clean greenhouse environment (Kumari, R., & Kumar, R. 2019).
4. A key element of aeroponic hydroponics is the misting system, which provides fine water droplets to plant roots. This system combines pipes with nozzles to spray nutrient solutions onto the roots straight from the Nutrient Mixer. This technique drastically lowers the water load because it requires a lot less water than conventional hydroponics. Although the system uses less water overall, it still consumes a lot of water due to nozzle clogging, nozzle stability, and uneven mist distribution across the roots, among other technical issues. Therefore, more study and innovation are needed to improve the Misting System's effectiveness and performance.
5. With its filtration mechanism that eliminates pollutants and root debris, the Drainage Manager system is an essential part of managing water that has traveled through plant roots (Higashide, T. et.al, 2004). Anaerobic bacteria can thrive in the reclaimed solution due to the presence of undigested organic compounds from the first misting operation. This can contaminate the solution and cause unpleasant aromas. This study uses a mix of a mechanical filter and a biofilter that can break down organic particles to address this problem. The removal of organic com-

pounds is successfully improved by this integrated filtering system, which lowers the levels of water contamination and offers a practical way to keep the irrigation solution pure.

Advantages of Aeroponics

Conventional agricultural practices, including geoponics, can have detrimental environmental implications. Aeroponics has arisen as a breakthrough strategy in agriculture, where the demand for efficient and ecological farming methods is constantly increasing. This soilless form of plant cultivation has numerous benefits that meet the issues presented by traditional farming practices. Aeroponics is altering the agricultural landscape through increased nutrient uptake and precise environmental management (Pranjal Singh, et.al,2024).

1.Increased Nutrient Absorption:

Aeroponics is a novel and effective method for delivering nutrients to plant roots. By suspending the roots in a nutrient-rich mist, the approach guarantees that the nutrient solution has direct and optimal contact with them. This thin mist enhances nutrient absorption, far exceeding the capabilities of typical soil-based farming. As a result, plants in aeroponic systems have faster growth rates and higher yields, making this technology a game changer in the drive for food security.

2.Faster Growth Cycles:

Aeroponic systems provide precise environmental control, leading to faster plant development cycles. Growers can improve plant environments by adjusting temperature, humidity, and light exposure to meet the unique needs of each species. This level of control enables year-round cultivation and the capacity to optimize growth factors for maximum yield. Shorter growing cycles boost yields while also allowing for more adaptability to shifting market needs and environmental conditions.

3.High Crop Yields:

Aeroponics has proven to considerably boost crop yields compared to standard agricultural methods. The combination of improved nutrient absorption, shorter growth cycles, and optimal environmental conditions leads in a more productive system. This greater efficiency is critical in fulfilling the growing worldwide demand for food. As the world's population continues to expand, aeroponics emerges as a possible alternative for producing more food in a restricted space, making it an important tool for tackling food security issues.

4.Resource Efficiency:

Aeroponics stands out from traditional farming approaches due to its resource efficiency. The closed-loop technology reduces water use, making it a sustainable solution for water-scarce places. The technology uses water more efficiently than soil-based systems by providing critical nutrients for plant growth through nutrient mist. This efficiency not only saves water, but it also decreases the environmental impact of traditional farming, where excess water can cause runoff and contamination.

5.Elimination of Soil Borne Diseases:

Aeroponics eliminates the requirement for soil as a growth medium, reducing the risk of soil-borne illnesses that harm crop health. Traditional farming practices can transmit illnesses through contaminated soil, resulting in lower yields and a greater need for pesticides. Aeroponics generates a sterile environment by suspending plant roots in air and supplying nutrients via mist, reducing the risk of crop illnesses. This soilless strategy not only improves plant health but also decreases the need for chemical interventions, which aligns with the principles of sustainable and organic agriculture.

6.Adaptability to Urban Farming:

Aeroponics has a distinct benefit in its applicability to urban situations. Arable land is becoming increasingly limited as the world's urbanization trend continues. Aeroponic systems, particularly those built for vertical farming, enable cultivation in limited urban areas. Aeroponics, which uses vertical space, underutilized buildings, or repurposed structures, allows urban farmers to cultivate crops in a controlled environment. This versatility not only tackles the issue of limited space, but also minimizes the carbon footprint of moving produce from rural to urban locations.

7.Reduced Land Footprint:

Traditional farming sometimes necessitates extensive land to raise crops. In contrast, aeroponics reduces the land footprint required for agricultural cultivation. Plants can be stacked vertically, eliminating the need for soil and resulting in larger yields with less space. This decrease in land use helps alleviate difficulties connected to deforestation and habitat degradation, supporting a more sustainable approach to agriculture that corresponds with environmental

8.Year Round Crop Production:

Aeroponics allows for year-round food production, eliminating the seasonal limits of traditional agricultural methods. The controlled environment enables farmers to regulate variables such as light and temperature to imitate perfect growing conditions regardless of external weather patterns. This year-round production capability is especially useful in areas with harsh climates or short growing seasons, assuring a consistent and uninterrupted food supply.

9.Minimized Dependency on Arable Soil:

Aeroponics eliminates reliance on arable soil for agricultural growing. This is especially important in areas where soil quality is low or degraded. Aeroponics, by eliminating the need for fertile soil, offers up new agricultural options in regions previously thought to be unsuited for traditional farming. This component is critical for expanding agricultural operations to marginal lands, hence improving the overall resilience of food production systems.

10.Potential for Automation and Precision Farming:

Aeroponic systems are technologically advanced, making them ideal for precision farming.

Automated nutrient delivery, climate control, and monitoring systems can be integrated into aeroponic setups to reduce cultivation labor requirements.

Precision farming optimizes resources by ensuring that each plant receives the exact amount of nutrients required for maximum growth. Aeroponics' technology integration enhances efficiency and distinguishes it as a cutting-edge method to agriculture.

Challenges and Mitigations of Aeroponic System

Although aeroponics has many benefits, it also has some drawbacks. Aeroponic systems, like other agricultural innovations, face challenges that must be overcome for general success and acceptance(Pranjal Singh, et.al,2024).

1.Initial Setup Costs:

Aeroponic systems require specific equipment such misting devices, nutrient delivery systems, and environmental control mechanisms. Small-scale farmers or those with low financial means may find the first investment prohibitively expensive. High setup expenses are sometimes a hindrance, deterring potential users from embracing this new production method. Advancements in technology, competition among suppliers, and government incentives for sustainable agriculture can lower initial startup costs. As the sector grows, economies of scale may play a part in making aeroponics more affordable.

2.Technical Expertise Required:

Implementing aeroponics requires technical competence in system design, nutrient formulation, and environmental management. Aeroponic farming requires specialized equipment and knowledge of plant physiology in a soilless environment, unlike traditional farming methods that rely on years of experience. Individuals without prior experience with hydroponics or controlled-environment agriculture may face a steep learning curve. Providing education, training, tools, and guidelines can help persons successfully adopt aeroponics. Collaboration among researchers, industry specialists, and educational institutions can lead to the creation of user-friendly systems and materials.

3.Risk of System Failures:

Aeroponic systems are prone to technical problems, which can harm crop health. Aeroponic systems' delicate equilibrium can be disrupted by equipment problems, power outages, or nutrition delivery errors. Disruptions in nutrient availability, wet roots, or environmental factors can harm plant health and cause crop losses.

4.Energy Consumption:

Aeroponic systems, particularly in controlled-environment agriculture, require significant energy to maintain optimal conditions. Aeroponic systems use artificial lighting, heating, and cooling to provide optimal conditions for plant growth. Using non-renewable or carbon-intensive energy sources can increase operational expenses and negatively harm the environment. Aeroponic systems can lower their energy footprint by utilizing energy-efficient technologies and renewable energy sources like solar and wind. Optimizing system designs to reduce energy consumption and investigating alternate lighting options can promote sustainable practices.

5.Vulnerability to External Factors:

Aeroponic systems are vulnerable to external variables including power outages, extreme weather, and supply chain delays affecting crucial components. Unexpected conditions, such natural disasters or geopolitical events, might affect the availability of vital supplies in aeroponic systems. Supply chain interruptions for nutritional solutions, equipment, or replacement parts might disrupt continuous operations.

6.Limited Crop Variety:

While aeroponics succeeds in some crop varieties, it may not be ideal for all sorts of plants. Crops with extensive root systems or specialized soil requirements may not flourish in aeroponic environments. Root crops, such as potatoes and carrots, may struggle to adapt to the suspended root environment, limiting their potential for aeroponic cultivation. Research on crop-specific adaptations and innovative aeroponic system designs can broaden the spectrum of plants that flourish in these conditions. Crop selection should take into account each species' unique requirements and adaptations.

7.Water-Based Pathogen Risks:

Aeroponics eliminates soil-borne infections, but increases the risk of water-borne pathogens harming plant roots. Aeroponic systems' nutrient-rich mist has the potential to harbor diseases that injure plant roots. Inadequate management of infections can harm crop health and threaten system sustainability.

8.Scale and Standardization Challenges:

Scaling up aeroponic systems for commercial production and establishing industry standards are challenging. Aeroponics has shown effective in controlled tests and small-scale operations, but achieving consistency and scalability on a larger commercial scale is challenging. Standardizing procedures, nutrient formulas, and system designs for many settings and crops is a continuous problem. Collaboration among researchers, industry stakeholders, and regulatory agencies is crucial for developing standardized methods and recommendations for aeroponic farming. Pilot programs, data-sharing initiatives, and continual feedback loops can help enhance and standardize aeroponic technologies.

9.Public Perception and Acceptance:

Aeroponics may have obstacles in gaining public acceptance, especially in areas where traditional agricultural practices are deeply rooted. Resistance from consumers, policymakers, and traditional farmers may prevent widespread use of aeroponics. Perceptions of the safety, quality, and taste of aeroponically cultivated products can impact market adoption.

10.Regulatory Frameworks:

Inadequate regulatory frameworks for aeroponic technologies can hinder widespread deployment. Current legislation may not fully address the specific elements of aeroponic farming. Ambiguous legal frameworks can cause confusion for farmers, investors, and stakeholders, hindering the expansion of the aeroponics business.

Maintenance to be Carried Over for Aeroponics

To take into account the aeroponic system, certain conditions must be met, including an effective, trouble-free operation, cheap cost, long-term high plant productivity, and regular, scheduled maintenance. The aeroponic system's functionality and protection against severe weather must also be guaranteed. Because of this, aeroponic culture is vulnerable to the growth chamber's harsh or mild climate. Nonetheless, the following lists some of the most important regular and preventative maintenance points:

1) Verify the condition of the power source
2) Replace the nutrient solution on schedule
3) Verify that the air pressure pumps are operating as intended
4) Verify that the nutrient reservoir tank is full
5) Examine any leaks in the drain line and nutrient delivery system
6) Verify that atomization nozzles are operating in appropriate circumstances, such as when they become clogged
7) Verify that the nutrition solution's chemical characteristics are within the appropriate range

8) Verify that the environmental parameters are within an appropriate range

Aeroponics and Smart Farming

Using AI and IoT along aeroponics are the proposed system, where Greenhouse and vertical farming are the two main categories of indoor farming. The primary methods used in vertical farming are hydroponics, aquaponics, aeroponics, and bioponics (Fasciolo, B. et.al, 2023).

Plant fertilizer is utilized in aeroponic systems. The aeroponic method exposes the plant roots to aerosol droplets containing the nutritional solution. Certain plants, including bromeliads and epiphytic orchids, live in aeroponic systems in the wild, gathering nutrients from mist through their root systems and leaves. High-pressure atomization is a widely used process that involves forcing a high-pressure nutritional solution through an opening, causing the liquid to condense into droplets. This process produces droplets with a diameter of 10-100μm. The aeroponic system is made up of pipes that supply or recover the nutrient solution and a tank with plant roots positioned close to nozzles. A thermally insulated chamber houses the entire system. The main actuators are as follows:

- pumps and nozzles that provide the nutrient solution to the roots
- LED lights that provide artificial lighting to the plants
- air conditioning system and fans to regulate air circulation, temperature, and humidity
- piping system for nutrient solution recovery

Figure 5. Modified aeroponics (Fasciolo, B. et.al, 2023)

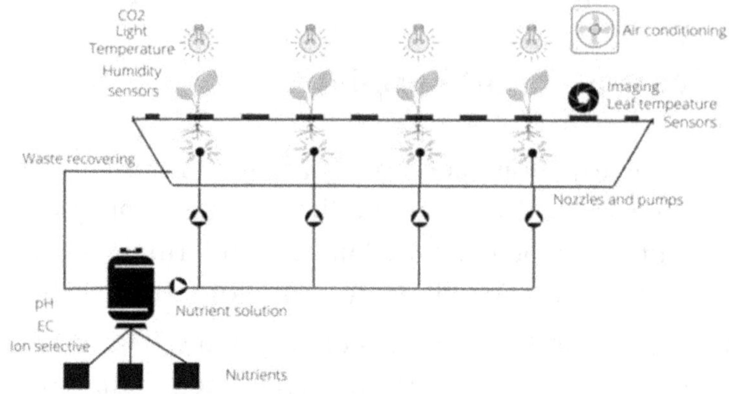

Aeroponics is believed to address a number of physiological issues with plants that arise with hydroponic farming. This may include improved water use efficiency and increased oxygen availability in the root bed. However, in comparison to more advanced hydroponics methods, it also necessitates a larger farm infrastructure and actuators. Furthermore, additional research is required on the nozzle type, drop size, irrigation frequency, and nutritional solution pressure (Fasciolo, B. et.al, 2023).

IoT and AI in Vertical Farming

Integrating IoT and AI with vertical farming is crucial to increasing its efficiency and ensuring that it has a less environmental impact. Sensors in an agricultural controlled environment are used by the system to gather data on plant growth and physiology. Artificial intelligence systems can then evaluate these data and monitor plant development through resource optimization. IoT has already been implemented and used in vertical farming systems, and in order to track plant growth, AI has recently been combined with IoT (Fasciolo, B. et.al, 2023).

Artificial intelligence (AI) in vertical farming often addresses two types of problems: prediction and control and optimization. In terms of forecasting, optimal harvesting times and precise crop yields have been predicted using models like Support Vector Regression (SVR). It has been demonstrated by several studies that an integrated method is preferable to utilizing SVM, Multiple Linear Regression (MLR), and Artificial Neural Networks (ANN) separately to estimate agricultural yields. Moreover, the threshold values of PH and other parameters have been predicted using Random Forest Regression. Regarding control and optimization, Random Forest Classification has been used to optimize the control function of grow lights, and Deep Neural Networks has been used to provide control action for a hydroponic environment in real time, such as controlling sprinkler and water flow based on PH, temperature, humidity, and water flowrate (Fasciolo, B. et.al, 2023).

One advantage of applying AI approaches is the reduction of production-related waste and pollution, as well as the removal of uncertainty connected to assessing plant health based on human opinion and more exact resource usage. However, it is also well known that AI has its limitations in situations where data is hard to get by or of low quality. Degraded sensors may produce inaccurate data, and weather, poor light, plant or insect obstruction, and foggy lenses can all reduce the quality of photos and videos taken by cameras and videos. Inaccurate monitoring data reports, protracted delays, and even data loss might come from unreliable sensing, processing, and transmission, which can ultimately affect the model's output (Fasciolo, B. et.al, 2023).

CONCLUSION

The global economy depends heavily on water. The agricultural sector uses over 70% of the fresh water that humans utilize. Out of that, extravagant irrigation methods waste forty-five percent of the crop. Because aeroponic systems use recirculatory systems, we can conserve 98% of the water used in total. Aeroponic systems can produce food quickly, efficiently, and with high quality all year round. It is possible to overcome any limitation associated with soil culture production with this soilless culture. India will soon rank among the top growers and exporters due to its increased production free of illness. Improved vegetative development may be achieved using an aeroponic system without the need for artificial hormones, insecticides, or pesticides. Even today, aeroponics is a valuable method for understanding plant requirements and cultivating plants in a controlled environment. There are situations where aeroponic gardening is the only feasible option for apartment dwellers who want to garden. Additionally, aeroponics solves this issue and offers the most efficient way to cultivate plants on arid lands.

REFERENCES

Ampim, P. A., Obeng, E., & Olvera-Gonzalez, E. (2022). Indoor Vegetable Production: An Alternative Approach to Increasing Cultivation. [Google Scholar] [CrossRef]. *Plants*, *11*(21), 2843. DOI: 10.3390/plants11212843 PMID: 36365296

Ashraf, M. A., Maah, M. J., & Yusoff, I. Soil Contamination, Risk Assessment and Remediation. In Environmental Risk Assessment of Soil Contamination; IntechOpen: London, UK, 2014. (Google Scholar) (CrossRef)

Benke, K., & Tomkins, B. (2017). Future Food-Production Systems: Vertical Farming and Controlled-Environment Agriculture. [Google Scholar] [CrossRef]. *Sustainability*, *13*(1), 13–26. DOI: 10.1080/15487733.2017.1394054

Campbell, C. Stories of the Year: What Does AeroFarms' Bankruptcy Signal for CEA's Future? The Food Institute. Available online: https://foodinstitute.com/focus/what-does-aerofarms-bankruptcy-signal-for-ceas-future/ (accessed on 26 December 2023).

Climate-Smart Agriculture. Available online: https://www.worldbank.org/en/topic/climate-smart-agriculture (accessed on 28 March 2024).

ElBeheiry, N., & Balog, R. S. (2023). Technologies Driving the Shift to Smart Farming: A Review. [Google Scholar] [CrossRef]. *IEEE Sensors Journal*, *23*(3), 1752–1769. DOI: 10.1109/JSEN.2022.3225183

Fasciolo, B., Awouda, A., Bruno, G., & Lombardi, F. (2023). A smart aeroponic system for sustainable indoor farming. *Procedia CIRP*, *116*, 636–641. DOI: 10.1016/j.procir.2023.02.107

Gnauer, C., Pichler, H., Schmittner, C., Tauber, M., Christl, K., Knapitsch, J., & Parapatits, M. (2020). A Recommendation for Suitable Technologies for an Indoor Farming Framework. EI Elektrotech. Inf. 137, 370-374.

Higashide, T., Shimaji, H., Hamamoto, H., & Takaichi, M. (2004). Higashide, T.; Shimaji, H.; Hamamoto, H.; Takaichi, M. A Method to Measure Low Drainage Flow and to Control Nutrient Solution Supply in Aeroponics Systems. [Google Scholar] [CrossRef]. *Seibutsu Kankyo Chosetsu*, *42*(4), 277–286. DOI: 10.2525/ecb1963.42.277

Karunathilake, E. M. B. M., Le, A. T., Heo, S., Chung, Y. S., & Mansoor, S. (2023). Karunathilake, E.M.B.M.; Le, A.T.; Heo, S.; Chung, Y.S.; Mansoor, S. The Path to Smart Farming: Innovations and Opportunities in Precision Agriculture. [Google Scholar] [CrossRef]. *Agriculture*, *13*(8), 1593. DOI: 10.3390/agriculture13081593

Knierim, A., Kernecker, M., Erdle, K., Kraus, T., Borges, F., & Wurbs, A. (2019). Smart Farming Technology Innovations—Insights and Reflections from the German Smart-AKIS Hub. [Google Scholar] [CrossRef]. *NJAS Wageningen Journal of Life Sciences*, 90–91, 100314.

Kumari, R., & Kumar, R. (2019). Aeroponics: A review on modern agriculture technology. *Indian Farmer*, *6*(4), 286–292.

Lakhiar, I. A., Gao, J., Syed, T. N., Chandio, F. A., Tunio, M. H., Ahmad, F., & Solangi, K. A. (2020). Overview of the aeroponic agriculture–An emerging technology for global food security. *International Journal of Agricultural and Biological Engineering*, *13*(1), 1–10. DOI: 10.25165/j.ijabe.20201301.5156

Lieder, S., & Schröter-Schlaack, C. (2021). Smart Farming Technologies in Arable Farming: Towards a Holistic Assessment of Opportunities and Risks. [Google Scholar] [CrossRef]. *Sustainability (Basel)*, *13*(12), 6783. DOI: 10.3390/su13126783

Nikolov, N. V., Atanasov, A. Z., Evstatiev, B. I., Vladut, V. N., & Biris, S. S. (2023). Design of a Small-Scale Hydroponic System for Indoor Farming of Leafy Vegetables. [Google Scholar] [CrossRef]. *Agriculture*, *13*(6), 1191. DOI: 10.3390/agriculture13061191

Pradhan, B., & Deo, B. (2019). Soilless farming–the next generation green revolution. *Current Science*, *116*(5), 728–732. DOI: 10.18520/cs/v116/i5/728-732

Redmond, K. Is Bankruptcy Filing Just a Temporary Setback for Aero-Farms? NJBIZ. Available online: https://njbiz.com/is-bankruptcy-filing-just-a-temporary-setback-for- aerofarms/ (accessed on 19 June 2023).

Roy, T., & George, K. J. (2020). Precision Farming: A Step Towards Sustainable, Climate-Smart Agriculture. In Global Climate Change: Resilient and Smart Agriculture; Venkatramanan, V., Shah, S., Prasad, R., Eds.; Springer: Berlin/Heidelberg, Germany, pp. 199–220. (Google Scholar) (CrossRef)

Streed, A., Kantar, M., Tomlinson, B., & Raghavan, B. How Sustainable Is the Smart Farm? In Proceedings of the LIMITS Workshop on Computing within Limits, Virtual, 14–15 June 2021. (Google Scholar) (CrossRef)

Wittmann, S., Jüttner, I., & Mempel, H. (2020). Indoor Farming Marjoram Production—Quality, Resource Efficiency, and Potential of Application. [Google Scholar] [CrossRef]. *Agronomy (Basel)*, *10*(11), 1769. DOI: 10.3390/agronomy10111769

Yépez-Ponce, D. F., Salcedo, J. V., Rosero-Montalvo, P. D., & Sanchis, J. (2023). Mobile Robotics in Smart Farming: Current Trends and Applications. [Google Scholar] [CrossRef] [PubMed]. *Frontiers in Artificial Intelligence*, *6*, 1213330. DOI: 10.3389/frai.2023.1213330 PMID: 37719082

Chapter 3
Importance of Aeroponics in Vegetable Crops

Wasan Salih Hussain
https://orcid.org/0000-0002-1537-9979
Mosul University, Iraq

Mahmoud Magdy Abbas
Independent Researcher, Egypt

ABSTRACT

Aeroponic means "growing in air." An aeroponic system is medium-less in that the roots of the plant are free-hanging inside an open root-zone atmosphere. Aeroponics structure supplies optimum levels of water, nutrients and air to the growing chamber. Aeroponics is the process of growing plants in an air or mist environment without use of soil or aggregate media. The word aeroponic is derived from the Latin word 'aero' (air) and 'ponic' means labor (work). This is an alternative method of soil-less culture in growth-controlled environments. The aeroponic culture technique is an optional device of soil-less culture in growth-controlled environments such as greenhouses. Aeroponics is more than just a farming method; it represents a paradigm shift in agriculture, offering a sustainable and efficient alternative to traditional farming practices. By embracing innovative techniques like aeroponics, we can work towards a more sustainable

DOI: 10.4018/979-8-3693-2320-5.ch003

and resilient agricultural system that meets the needs of growing global population while preserving our precious water resources

INTRODUCTION

Aeroponic means "growing in air." An aeroponic system is medium-less in that the roots of the plant are free-hanging inside an open root-zone atmosphere. Aeroponics structure supplies optimum levels of water, nutrients and air to the growing chamber. Aeroponics is the process of growing plants in an air or mist environment without use of soil or aggregate media. The word aeroponic is derived from the Latin word 'aero' (air) and 'ponic' means labor (work). This is an alternative method of soil-less culture in growth-controlled environments. The aeroponic culture technique is an optional device of soil-less culture in growth-controlled environments such as greenhouses.

Aeroponics is more than just a farming method; it represents a paradigm shift in agriculture, offering a sustainable and efficient alternative to traditional farming practices. By embracing innovative techniques like aeroponics, we can work towards a more sustainable and resilient agricultural system that meets the needs of growing global population while preserving our precious water resources

Aeroponics is a way of planting in which plants are suspended in the air and grow in a humid environment without soil. This technique has the most advantages compared to other planting techniques in different parts of the world. With aeroponics, one can control humidity, temperature, pH, and water conductivity under a greenhouse.

Aeroponics can be applied in many different ways. The principles of aeroponics are based on the possibility of cultivating vegetables whose roots are not inserted in soil but in containers filled with flowing plant nutrition. In these containers' roots can find the best condition regarding oxygenation and moisture. These conditions allow for better plant nutrition assimilation in a more balanced way, with consequential faster development of the cultivated plants. The aeroponic system is more user-friendly as the plants are all separated, they are all suspended in the air and the roots of the plants are not in anything like soil or water. Also, the harvesting of crops is simple. Many vegetable crops like potatoes, yams, tomatoes, lettuce, and some leafy vegetables are being commercially cultivated in an aeroponic system. Aeroponic system has been successfully for potato seed tuber production. Aeroponic systems are more water resource efficient than hydroponic system. Another remarkable advantage of the aeroponics is the minimal contact between the support structure and plant, due to which the unconstrained growth of the plant is possible. The aeroponics systems are widely used for NASA space research programs. Because it is the best technology for agriculture outside the Earth's system, aeroponic agriculture does not depend on the planet's gravity to move the nutrient solution when planting in soil or hydroponics, but rather the solution is flown by a pressure pump; therefore, agriculture can be done in space where there is no gravity.

In order for plants to grow, they need air from which they take oxygen and carbon dioxide, and they also need water, nutrients, light, heat, and mechanical fixation. In traditional soil agriculture, the plant obtains mechanical fixation, nutrients, and water from the soil. Plant nutrients can be classified into two types: major nutrients and minor nutrients; major nutrients are those that the plant needs in abundance, such as: oxygen, hydrogen, and carbon that plants get from water and air, and the major primary nutrients are nitrogen, phosphorus, and potassium, and the major secondary nutrients are calcium, sulfur, and manganese. As for the minor nutrients, they are those that the plant needs less, and they are: boron, chlorine, manganese, iron, zinc, copper, molybdenum, and nickel.

AEROPONIC AGRICULTURE

Global climate change has increased the risk of recurrent droughts. Agriculture is undergoing a major change around the world and is facing serious problems. It is expected that in the future, it will be difficult to provide a new and clean food supply for the rapidly growing population using traditional agriculture. These reasons have prompted researchers and scientists to think of innovative ways, namely soilless agriculture, which is an alternative technology to achieve effective adaptation. The soilless system is the hydroponic system and aeroponics. In the aeroponics system, the roots of plants are suspended in a plastic carrier and foam, replacing the physical soil under controlled conditions. The roots are allowed to hang freely and openly in the air. However, the nutrient-rich water provides the nozzles with different droplet sizes intermittently or continuously. The nozzles create a fine mist, which is the best way to grow plants for food security and sustainable development and the most efficient, beneficial, important, economical and convenient.

History

Aeroponics system was invented in Italy, University of Pia, by Dr. Franco Massantini. Aeroponics was developed in 1942 by Carter .W. The aim was to facilitate the examination of plant roots. Between the 1940s and 1950s, scientists used aeroponics in the cultivation of tomatoes, apples, citrus fruits, etc. in order to observe the structure of their roots and test their resistance threshold to different types of rot resulting from over-watering or severe drought. The system can be vertical, horizontal or both. Pipes and containers made of polyvinyl chloride (C.V.P.) are used, in the vertical position, filled with holes to insert the plants, so the roots grow inside the cylinder, and are completely dark and exposed to air and no soil, but suspended in the medium that supports them. The plants start from cuttings or seeds, then suspended in the air in a chamber called the root chamber(Buckseth, 2016). The root system of the plants thrives in a closed space, where a solution rich in nutrients is regularly sprayed.

Types of Aeroponic Farming System

1. **Low-pressure aeroponics**: This basic type of aeroponic system uses a low-pressure pump to provide a nutrient solution to the plants growing through jets or ultrasonic transducers. The system also allows drips that take nutrients back to the reservoir Fig(1).
2. **High-pressure aeroponics**: In this system the nutrient solution to the roots at high pressure, with the help of a high-pressure pump and specialized misting nozzles. High-pressure aeroponics allows for more

precise control over the size of the mist droplets, which can lead to better nutrient absorption by the roots.

3. **Vertical aeroponics**: In these systems, Plants can be grown in stacked layers because of their design, which maximizes the use of vertical space.
4. **Aeroponic towers**: Aeroponic towers allow for natural plant growth while conserving 95% of the water needed for conventional organic farming. This setup is composed of tall, vertical structures which can accommodate multiple growing trays or channels. This type of aeroponics farming system is currently gaining popularity in urban and peri-urban areas

Figure 1. High pressure vs low-pressure aeroponics

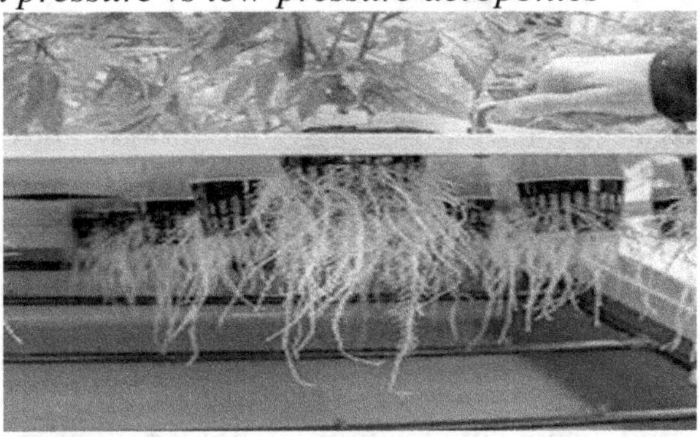

Innovations in Aeroponics

The role of technology in aeroponics is evident at all stages of its life cycle, including careful moderation of environmental factors such as water conservation, space optimization, and pest and disease control.

To facilitate our discussions, we have divided the role of technology in aeroponics into four sectors: sensors to monitor the environment, the use of the Internet of Things, data, and artificial intelligence to make the production process smarter, technology to improve dispensers and renewable energy technologies (Kumari et al.,2019).

Environmental Monitoring Sensors

These sensors monitor a range of environmental factors in real-time, including temperature, humidity, and nutrient levels. They enable farmers to make informed decisions on how to optimize their operation. Additionally, these solutions come with automation system integration options to adapt to changing environmental conditions.

1. Using AI(Artificial Intelligence), IoT, and Data

These solutions, often referred to as Industry 4.0, help improve efficiency in multiple ways. They can automate repetitive tasks, detect plant diseases early, suggest appropriate preventive measures, and analyze large amounts of data critical to plant growth.

2. Technology to Improve Dispensers

Dispenser-related technologies play a pivotal role in the success of aeroponics. The function of aeroponics is to deliver nutrients and water to the plants. These dispensers include atomizers, sprayers, foggers, ultrasonic dispersers, water pumps, and centrifugal pumps. Having the best-in-class dispensing technology ensures that the plants get the exact dosage. Automated dispensing technology helps reduce manual labor and saves resources.

3. Renewable Energy Technologies

Aeroponics is a suitable agricultural technology to take advantage of multiple alternative and renewable energy options. One can power aeroponics farms through solar or wind energy sources, reducing reliance on environmentally harmful grid power. Judicious use of renewable energy can also be very cost-effective for aeroponics farmers when they achieve large scale. All these technological interventions should make optimum contributions to making aeroponics an effective farming of choice.

However, since aeroponics is an emerging technique, and is still growing in its maneuvers, it will take some time to maximize the returns on each of these investments.

Aeroponics: Opportunities and Challenges

In terms of opportunities, aeroponics requires moderate infrastructure. All you need is a light source to enable the plants to photosynthesize and store energy, a cloth-like medium to separate the plants from their roots and give them support to grow, and a nutrient-dense mist in the solution chamber for the root systems to absorb.

Since aquaponics recycles the nutrient solution, it doesn't waste water. However, it still requires electricity to power the pumps and artificial light to help the plants grow.

For aeroponics to truly thrive, training and education of growers is essential. Growers need to continually improve their skills to keep up with a field that is thriving with innovation and new technologies. One area where training is essential is how to keep the system clean.

In general, success in aeroponics depends on successfully dealing with a range of factors. Improper handling of one can bring down the entire system.

Aeroponics must also have a robust structure that allows for adequate light and air supply, especially to exposed plant parts.

Aeroponics, which works through two main components, the roots and the canopy, is also partly based on Aeroponics. Therefore, any grower keen on taking advantage of hydroponics must have the right knowledge of hydroponics. For example, aeroponics must have a backup system to provide essential nutrients to the plants in case the current system fails.

A high-pressure diaphragm pump (80 psi) is used in high-pressure aeroponics to supply nutrients to the roots through 20-50 mm mist heads. Such emergency knowledge is essential to successfully run an aeroponics farm.

This can only come through thorough training and education on how to grow different plants and crops under diverse conditions.

The United Nations expects the world population to rise to 9.4-10.1 billion people in 2050. Currently, only about 3 percent of the Earth's surface is arable land. This population growth is accompanied by the need to produce more food to feed the growing population, with the decreasing areas of arable land and some of it being out of service due to poor land management, salinity, diseases, and the continued expansion of cities at the expense of arable areas. The challenge of producing more food in a sustainable manner will become more urgent than ever before, as work is underway to produce more productive seeds that are tolerant of climate conditions and resistant to diseases. Countries are also urging farmers to adopt modern agricultural techniques to overcome these obstacles, and one of the most important of these modern agricultural techniques is soilless agriculture (Al-Quraan, 2023).

How Does Aeroponics Contribute to Enhancing Food Security?

Food and water security are at the top of the priorities of Arab countries that suffer from a severe shortage of water resources and arable land, in addition to dry weather and very high temperatures.

These countries consume large amounts of food resources annually.

Due to their desert nature, the percentage of desert and rocky desert areas reached 41.89% of the total area of Iraq, while water areas decreased by 1.25% of the land area, and the Tigris and Euphrates rivers constitute the most important sources of these areas, in addition to the scarcity of fresh water resources.

Statistics also showed that abandoned agricultural lands constitute the second largest area at 22.21%, reflecting the challenges facing farmers in such environments, such as water scarcity, deterioration of soil health, and lack of agricultural inputs and support (Central Statistical Organization, 2024).

The deterioration of the agricultural situation in Iraq, and the resulting shortage in the availability of food materials and its serious repercussions on national food security, provided a strong incentive to research the issue of food security and what it means in terms of the role of bread and the importance of food in the lives and security of peoples, especially after the huge increases in population, and the accompanying crises in food materials, due to environmental and climatic changes and the deterioration of political conditions in many countries, all of which put many countries, including Iraq, in the circle of risk of famine and humanitarian disasters (Al-Dhalimi, 2009).

Among the innovative technologies that can be used to enhance food security are aeroponic farms, which are mobile and inflatable farms.

Among the innovative technologies that can be used to enhance food security are wind farms, which are mobile and inflatable farms.

The project integrates more than 15 innovative technical solutions, including the Internet of Things and artificial intelligence to create a sustainable environment for food security and agriculture.

Advanced smart agriculture solutions enhance crop productivity by high rates, in addition to reducing labor costs and reducing the need for agricultural supplies such as water and fertilizers.

Benefits of Aeroponic Cultivation of Vegetables

Global climate change has increased the risk of recurrent drought. Agriculture is in a phase of major change around the world and dealing with serious problems. It is expected that in the future, it will be difficult to provide new and clean food supplies for the rapidly growing population using traditional agriculture. These reasons have prompted researchers and scientists to think of innovative ways, namely soilless agriculture, which is an alternative technology to achieve effective adaptation. The soilless system is the hydroponic system and aeroponics. In the aeroponics system, the roots of plants are suspended in a plastic carrier and foam, replacing the physical soil under controlled conditions. The roots are allowed to hang freely in the air. However, the nutrient-rich water provides the nozzles that create a mist of different droplet sizes intermittently or continuously. The nozzles are the best way to grow plants accurately for food security and sustainable development and the most efficient, beneficial, important, economical and convenient (Garzon, et al., 2023).

Many crops can be grown in such a system, especially leafy vegetables, such as: mint, basil, parsley, radish, arugula, lettuce, mallow, spinach, cabbage, and other vegetables such as: cucumber, tomatoes, hot peppers, sweet peppers, cauliflower, and other vegetables. Legumes can also be grown, such as: beans, okra, beans, beans, and others. Some fruits can be grown such as: strawberries and cantaloupe, and even larger plants such as: bananas, apples, and lemons if they are provided with mechanical support; and one system like this can provide self-sufficiency in salad daily for the home.

There are many benefits to aeroponic cultivation compared to soil cultivation and hydroponics, as aeroponic cultivation requires very small spaces; hundreds of plants can be grown in an area not exceeding 10 square meters, as aeroponic cultivation is very suitable for vertical farming. Ready-made towers are now sold for aeroponic cultivation that can be placed in the home garden, and there are even types that can be placed in a closed room, as these towers contain high-power photodiodes that provide plants with enough light for photosynthesis. Aeroponics saves large amounts of water and fertilizers, exceeding 95% compared to soil cultivation, and its water use efficiency reaches nearly 100%, as the only water lost in this method is what evaporates from the plants and what is stored in them; because the water in aeroponic systems is recycled. Plants grown in the air grow faster than if they were grown in soil or even in water, and are clean and less susceptible to insects and pests, and do not emit unpleasant odors. The reason for the rapid growth of plants is that the latter obtain the nutrients they need with ease; and since the roots are suspended in the air, they obtain oxygen in larger quantities than other types of agriculture, and the roots are less susceptible to rot and diseases. Aeroponics requires less care, and can be programmed to be fully automated, as it does not require fertilization, plowing, or any operations in the soil. One of the advantages of aeroponics is that it works almost anywhere, so even if there are no spaces in the home garden, it can be set up on the roof of the house. It can also be grown inside rooms as soon as it is provided with appropriate lighting. Aeroponics are now available in the market at relatively cheap prices.

Aeroponics is one of the most rapid propagation methods useful in vegetable production. The technique can produce large numbers of mini plants in one generation, thus eliminating the need for more field multiplications reducing costs, and saving time. One study found that aeroponics can produce over 100 mini tubers of an individual potato plant in a single row (Otazu, 2008) as opposed to the conventional method that creates approximately 8 daughter tubers only in a year while only 5–6.

This method, often implemented in greenhouses for optimal environmental control, offers several advantages, including increased crop yields, precise control over plant nutrition, and a reduction in water usage. Moreover, aeroponics eliminates the need for pesticides, resulting in a cleaner and more sustainable form of agriculture.

aeroponics vegetable crops represent an advancement in artificial life for plant growth and seed germination support systems, environmental control, and techniques of drip irrigation traditionally used in agriculture and its suitability for zero gravity environments compared with conventional methods, aeroponics can significantly increase crop yields by 45% to 75%, while using water more efficiently, with nearly 99% of water utilized (Bodduet.al.,2024).

Aeroponics does not require soil fertilizers or pesticides, so the fruits and vegetables produced can be pure and do not require washing before consumption. Nutrients' Direct delivery to plant roots promotes faster crop growth and healthy yields.

This system is successful more than ten times more than conventional techniques like tissue culture and hydroponics, while also conserving water and energy through nutrient solution recirculation.

Aeroponics has been shown to improve root growth, survival rates, growth rates, and maturation times. It can also improve tuber yield and root aeration by suspending plants in the air.

Table 1. Some of the benefits and limitations associated with aeroponic system(Kaushik, and Tare, 2024)

Benefits	Limitations and Challenges
Water Efficiency	High Initial Cost
Faster Growth Rates	system Complexity
High Yields	Risk of System Failure.
Space Efficiency	Sensitivity to Power Outages
Nutrient Control	Nutrient Imbalance
Reduced Pests and Diseases	Heavy Reliance on Artificial Lighting system
Higher Efficiency.	Risk for contamination

CONCLUSION

It can be concluded that of the various systems adopted for the production of pre-basic seeds of potato, the aeroponic system appears to be the best in many respects. Considering the potential benefits of the system such as rapid production of seed, spacious, good nutrient monitoring system, improvement of growth and survival rate of plantlets, constant air circulation and ecologically friendly, this system has a potential of revolutionizing potato seed production industry. This system not only

REFERENCES

Al-Dhalimi, R. B. S. (2009). *Food Security in Iraq: Problems and Solutions*. Badia and Lake Sawa Studies Center.

Al-Quraan, I. M. (2023). *Practical applications in soilless agriculture and hydroponics*. Food and Agriculture Organization of the United Nations.

Boddu, V., Kumar, D. H., Kumar, S. N., Bindiya, Y., & Rajani, A. (2024). Aeroponics in Vegetable Crops. *International Journal of Theoretical & Applied Sciences.*, *16*(1), 74–78.

Buckseth, T., Sharma, A. K., Pandey, K. K., Singh, B. P., & Muthuraj, R. (2016). Methods of pre-basic seed potato production with special reference to aeroponics—A review. *Scientia Horticulturae*, *204*, 79–87. DOI: 10.1016/j.scienta.2016.03.041

Garzon, J., Montes, L., Garzon, J., & Lampropoulos, G. (2023). Systematic Review of Technology in Aeroponics: Introducing the Technology Adoption and Integration in Sustainable Agriculture Model. *Agronomy (Basel)*, *13*(10), 2517. DOI: 10.3390/agronomy13102517

Kaushik, K., & Tare, K. (2024). Aeroponics Farming System: A Step Towards Modern Agriculture. *Food and Scientific Reports*, *5*(5), 48–52.

Kumari, R., & Kumar, R. (2019). Aeroponics: A Review on Modern Agriculture Technology. *Indian Farmer.*, *6*(4), 286–292.

Chapter 4
Working Method of Aeroponics: Innovations in Soilless Agriculture

Froilan Delute Mobo
https://orcid.org/0000-0002-4531-8106
Philippine Merchant Marine Academy, Philippines

ABSTRACT

The chapter "Working Method of Aeroponics: The article "Innovations in Soilless Agriculture" explains how aeroponic farming works, and the parts of an aeroponic system, with particular concentration on particular farming procedures. Aeroponics provides an organic technique to address contemporary horticulture's problems by delivering suspended plant roots in air with nutrient enriched mist. Its advantages include better nutrient uptake, good aeration and very reduced water and other resource use as compared to conventional soil-based farming. Each of them is elaborated on in the chapter and described its influence on the growth of plants in the given aeroponic system; these components include the growth chamber, the system of misting the leaves, nutrient supplying, and the environmental factors regulating the plant growth.

DOI: 10.4018/979-8-3693-2320-5.ch004

INTRODUCTION

Thus, AE can be seen as a radical shift in traditional agricultural practices since it is possible to consider aeroponics as a kind of groundless farming. This system bathes plant roots in the air and provides nutrients through droplets, ensuring that the plant is exposed to oxygen all the time so that it grows at a very fast rate. The aeroponic method proves to be more appropriate as a solution to the existing problems of the world, for example, food shortages and population growth, as well as the lack of resources. Housing an explanation of the principles, components, and operational techniques of aeroponics, this chapter seeks to stress the transformative nature of aeroponics in present-day agriculture (Birkby, 2016).

Principles of Aeroponics

Aeroponics is a system of high-soil-less cultivation where plants are suspended in the air with their roots exposed to air and/or mist. The essence of aeroponics is based on the direct supply of nutrients to the plant roots, which are suspended in the air, thus exposing them to oxygen and achieving maximal nutrient absorption (Kyriacou & Rouphael, 2018). This is even different from hydroponics, where plants are grown in water, and soil-less farming, which involves the use of soil for anchorage as well as for nutrient provision.

In other words, the simple impotent of aeroponics is spraying the plant roots with a nutrient solution at certain intervals. This way, it is possible to satisfy the needs of roots, providing them with necessary nutrients and, at the same time, necessary oxygen. With the use of hydroponics, various problems associated with soil are negated, for instance, soil-borne diseases, weeds, and pests, all of which affect crop production and quality of produce (Kozai et al., 2016).

Compared to traditional soil-based farming, aeroponics gives young plants control over nutrient content, pH, and moisture. mainly because of this high level of pest control, which in turn increases the growth rates, yield, and quality of the produce (Despommier, 2011). Besides, the aeroponics require comparatively a small amount of water, reaching up to 5%

of water compared to conventional farming methods; this is beneficial in areas with weak water access (Birkby, 2016).

An average aeroponic system has the following parts and subparts: in addition to the minute structures and parts that make up the aeroponic systems that enable a plant to grow healthy.

The growth chamber is where the plant roots are mounted and allowed to be sprayed with the nutrient mist. The design of the chamber is important to avoid exposure to light from roots since it is detrimental to them and to ensure the provision of the right humidity levels for the proper growth of the plants. In the same regard, the growth chamber should have the appropriate drainage system in order to avert water logging as well as to ensure that oxygen gets to the roots at all times (Kyriacou and Rouphael 2018).

Misting System

The misting system is the central fixture in aeroponics. It employs high-pressure pumps and very fine mist-producing nozzles that deliver a nutritional mist that has not been waterlogged and is therefore fine enough to stick to plant roots. This is done by regulating the number of times and the period within which the plants are misted so that the right amount of nutrients and moisture are delivered to the roots without the plants getting dry or oversaturated (Kozai et al., 2016).

Nutrient Delivery Systems

The nutrient delivery system in an aeroponic system involved the preparation and circulation of nutrient solutions. It comprises all the basic elements required for the growth of plants like nitrogen, phosphorus, potassium and the micronutrients. The delivery system is unique in that it cuts down on nutrient loss through the cycling of the solution, thus having low operational costs (Despommier, 2011).

Environmental Control Systems

Facilities such as environmental control systems are critical in as far as they relate to providing the best growing conditions. These systems control some factors, such as the temperature, humidity, concentration of CO2, and light intensity within the growth chamber. These parameters can be regulated in the current sophisticated aeroponic systems by using sensors and automation to 'notify' when modifications to the environment are required to enhance plant growth (Kozai et al., 2016).

Automation and Monitoring Systems

There are many components when it comes to aeroponic systems, and automation is one of the most essential because the system's success relies on the method and frequency through which nutrients and mist are to be delivered and the conditions under which the crops are to be grown. HCO3, SO4, NO3, and PO4 electrical conductivity and temperature can be measured by sensors for modifying or knowing the status of the environment within which the crop is growing (Kyriacou & Rouphael, 2018). This brings down the costs of a workforce, and the reliability and efficiency of plant tendencies increase.

WORKING PRINCIPLES OF AEROPONICS

The working principles of aeroponics concentrate on the efficiency of the nutrients in the nutrient solution, oxygen uptake, and the environment in which the plants are grown.

Nutrient Absorption

Aeroponic systems feed nutrients to the root systems of plants in a very easy-to-use-to-use form of nutrition. The misting process makes it possible for the nutrients to be delivered to the entire surface of the roots and, hence, absorbed by the plant as quickly as possible. Another advantage of vertical farming is that it generally lacks soil; this means that there is less potential for the phenomenon of nutrient lockout that can occur in ordinary soil, where certain nutrients become sequestered by the soil chemistry (Birkby, 2016).

Oxygenation of Roots

Another great benefit of aeroponics is the possibility of providing the plant roots with a large amount of oxygen. Bryophytes that have their roots hanging in midair have access to oxygen that they need to metabolize and absorb nutrients. Consequently, improved oxygen supply to the roots develops exuberant root development and minimizes the prospects of root ailments, including root rot, which is normal in anaerobic or waterlogged ground (Kozai et al., 2016).

Water and Nutrient Conservation

The system of growing crops aeroponically is very economical in terms of consumption of water and nutrients. The nutrient solution is recirculated, and miles of piping mean that waste is minimized. Research has identified that in the application of aeroponics, water use could be cut out by up to 95%, something that could make the system desirable in areas where water scarcity is a big issue (Kyriacou & Rouphael, 2018).

Growth Optimization

In aeroponics, therefore, growth optimization is realized by constant monitoring of parameters such as nutrient delivery and growth conditions. This enables the growers to achieve maximum growing conditions for the crops that they are growing, resulting in high yields and high-quality produce (Despommier, 2011).

Aeroponics is advantageous in that aeroponics has the following disadvantages:

Aeroponics has several advantages, yet it also has a number of limitations that need to be discussed in order to understand if it is worth applying this method.

Advantages

- **Increased Growth Rates:** Because of the high oxygen concentrations and better nutrient-supplying capacity aeroponic systems enable, generally plants have better growth rates in comparison to soil (Kyriacou & Rouphael, 2018).
- **Higher Yields:** More efficient growing environments and lower disease pressure mean higher productions and improved quality of produce (Kozai et al., 2016).
- **Water and Nutrient Efficiency:** Aeroponics is portable, easy to mechanize, requires little water, uses up to 5% less water than conventional agriculture, and loses minimal nutrients through recycling.
- **Space Utilization:** Vertical systems of aeroponics can also be done so that they do not take a lot of space and are perfect for the growing of crops in an urban setting and other places where space is a limiting factor (Despommier, 2011).

Disadvantages

- **High Initial Costs:** Aeroponic systems could be expensive to install because specialized equipment such as high-pressure pumps, mist nozzles, and environmental controls must be installed (Kyriacou & Rouphael, 2018).
- **Technical Complexity:** Aeroponic systems are rather complex and costly, and their implementation necessarily involves the application of high technical knowledge that may be a challenge to the growers, particularly those who are not ready-made, using modern high-tech agricultural equipment (Birkby, 2016).
- **Dependency on Power Supply:** Aeroponic systems depend on electricity, which is used to run the pumps and other essential things for controlling the environment. The interference of the power source may lead to the downfall of the system, while the crops may also be affected (Kozai et al., 2016).

APPLICATIONS AND FUTURE PROSPECTS

Aeroponics is very much in demand at the present time and is expected to continue to be in high demand in the future.

Current Applications

- **Commercial Agriculture:** Aeroponics is used in commercial farming to cultivate crops such as vegetable sprouts, herbs, and microgreens. These crops gain from the fast growth rates and quality produce associated with aeroponic production (Despommier, 2011).
- **Urban and Vertical Farming:** Its compactness makes aeroponic systems suitable for applications where space is limited, such as in urban and vertical farming. Aeroponics also encourages plant density; numerous plants are packed within a small area; in this case, yield per unit area is enhanced, especially if practiced in urban areas (Kyriacou & Rouphael, 2018).
- **Research and Development:** Aeroponics, as already mentioned, is commonly used in plant research for the following reasons: root growth analysis, nutrient adsorption, and measurements under well-defined environmental conditions. There is always the option of experimenting with new crop varieties and coming up with new techniques in farming (Birkby, 2016).

Future Prospects

- **Space Agriculture:** Thus, the technique of aeroponics has been considered to supply the necessity of food production in space. Because of low water and nutrient consumption and the possibility of functioning in a closed system, it is appropriate for long-term space missions (Kozai et al., 2016).
- **Technological Advancements:** The upcoming automation and sensor improvements are expected to enhance the efficiency and scalability of aeroponic systems. This could be a result of the incorporation of artificial intelligence and machine learning to enhance the control of growing conditions, enhance yields, and cut costs (Kyriacou & Rouphael, 2018).

- **Sustainable Agriculture:** The world population is increasing in equal measure, and this has put pressure on the world's agricultural systems to produce a lot of food using scarce resources. Aeroponics can cater to this challenge by using less water and nutrients and, at the same time, producing numerous yields (Despommier, 2011).

CONCLUSION

Aeroponics is one of the most promising types of soilless agriculture, which has a number of advantages compared to traditional methods of planting. Thanks to its capacity for efficient delivery of nutrients and oxygen, low evaporation and loss of water and nutrients, and effective generations of crops of high quality in controlled circumstances, hydroponics could be a good solution to counter the difficulties of today's methods of farming. On the other hand, the initial costs and technicalities of the aeroponic systems can be relatively high, which is a hindrance to their promotion. New technological developments and innovations, together with emergent trends in environmentally sustainable farming of crops, are likely to enhance the scale of aeroponic production in the coming years.

REFERENCES

Birkby, J. (2016). Vertical farming. ATTRA Sustainable Agriculture. https://attra.ncat.org/attra-pub/vertical-farming/

Despommier, D. (2011). *The vertical farm: Feeding the world in the 21st century*. Picador.

Kozai, T., Niu, G., & Takagaki, M. (2016). *Plant factory: An indoor vertical farming system for efficient quality food production*. Academic Press.

Kyriacou, M. C., & Rouphael, Y. (2018). Towards a new definition of quality for fresh fruits and vegetables. *Scientia Horticulturae, 240*, 36–42. DOI: 10.1016/j.scienta.2017.09.046

Chapter 5
Aeroponics Techniques for Improved Farming Using Artificial and Deep Learning techniques

C. Amuthadevi

https://orcid.org/0009-0008-4020-918X

Department of Computational Intelligence, SRM Institute of Science and Technology, Kattankulathur, India

E. Afreen Banu

https://orcid.org/0000-0001-8654-9851

Department of Computer Engineering, Shah and Anchor Kutchhi Engineering College, Mumbai, India

S. Sampath Kumar

https://orcid.org/0000-0002-5368-3357

Computer Science and Engineering, Sri Eshwar College of Engineering, Coimbatore, India

S. Karthick

https://orcid.org/0000-0002-6563-0172

Department of Electronics and Communication Engineering, Erode Sengunthar Engineering College, Erode, India

A. Kistan

https://orcid.org/0000-0003-1334-4331

Department of Chemistry, Panimalar Engineering College, Chennai, India

DOI: 10.4018/979-8-3693-2320-5.ch005

M. Sudhakar
https://orcid.org/0009-0000-3045-3834
Department of Mechanical Engineering, Sri Sai Ram Engineering College, Chennai, India

ABSTRACT

This chapter will delve into the integration of aeroponics into artificial intelligence and deep learning techniques for agricultural productivity and its sustainability. This is a no-soil farming method where plants grow in a nutrient-rich mist. Therein lies a couple of major advantages: water efficiency and an accelerated pace of plant growth. The reasoning behind the inclusion of AI and deep learning techniques lies in the computer vision and predictive analytics ability—how best it can help determine if this has the potential to optimally run aeroponic systems. Monitoring the health of a plant and the environmental state in real time, using AI-driven sensors and deep learning for analytics, is capable of identifying data patterns in predicting growth and optimizing practices for the delivery of nutrients. Some specific successful cases and novel innovations are exemplified next, showing how new breakthroughs can solve the existing challenges of agriculture, improve yield quality, and ultimately reduce resource consumption.

INTRODUCTION

Aeroponics is the redefined and advanced method of agriculture, which grows plants sans the use of soil. Instead, the plants are suspended in the air and fed with a fine mist of nutrients. Hence, it causes the plants to grow faster and makes them consume fewer resources. Many advantages associated with this practice include efficient usage of water, very high availability of oxygen to roots, and less risk of diseases compared to the conventional method of growing in soil. However, fully bringing out such benefits to the fore, and eliminating the bottlenecks for widescale adoption of aeroponics, call for increasing integration with advanced technologies in AI and deep learning (Mehra et al., 2018).

Artificial intelligence and deep learning do not bypass any industry, and agriculture makes no exception. These technologies provide for the development of intelligent systems capable of analyzing huge amounts of data, making decisions in real-time, and optimizing processes. In the context of aeroponics, AI and deep learning are capable of improving the precision and efficiency of plant cultivation to a great extent. AI-driven sensors and deep learning algorithms will make it possible to trace all environmental variables, track plants' health, and work on delivering systems of nutrients more effectively than ever (Narimani et al., 2021).

This involves setting up aeroponics systems equipped with multiple sensors to constantly acquire data regarding temperature, humidity, light intensity, and nutrient concentration. The output from these sensors will, in turn, develop data streams in real-time processed and analyzed by AI algorithms. In this regard, deep learning models—especially using convolutional neural networks (CNN)—analyze images of plant leaves to detect very early signs of disease or nutrient deficiency. This ensures timely interventions and reduces the risk of crop failure by enhancing overall plant health (Torres-Tello & Ko, 2023).

AI-powered systems optimize nutrient delivery after analyzing big data related to the growth patterns of plants and environmental conditions. Predictive analytics, a subset of AI, uses historic data and current performance to project future demands of nutrient needs. This readjustment of the nutrient solutions dynamically ensures that plants receive the optimum balance of the essential elements for their growth, which promotes robust growth and maximum yields (Rajendiran & Rethnaraj, 2023).

One of the major challenges facing aeroponics is maintaining the proper balance of nutrient supply with environmental factors. AI can help surmount this challenge through actionable insights and recommendations. Machine learning models, for instance, might predict how changes in temperature or humidity affect plant growth and give recommendations to the aeroponic system. Such proactive action helps to avoid a lot of common problems like nutrient imbalances and environmental stressors that may adversely affect yields, thus assuring high and more uniform quality yields (Gowtham & Jebakumar, 2023).

Other advantages of incorporating AI and deep learning into aeroponics are autonomous methods of managing the system. AI algorithms would take control of misting systems, lighting, and other environmental factors with real-time data; adjustments do not have to be made manually, which could maybe reduce human-made mistakes to a minimum. Automation ensures smooth processes and uniform care for the plants to ensure better performance and yield. Increased integration of AI and deep learning in aeroponics also offers an opportunity for sophisticated research and development (Fasciolo et al., 2023). Artificial intelligence tools could model a range of growing conditions and their influences on plant growth. In this way, new cultivation strategies can be evaluated, and more efficient kinds of aeroponic systems can be developed. With AI-driven insight, the development of new lines of plants propagated for aeroponics and much more will be realized.

AI and deep learning in aeroponics is a relatively new area of research. Thus, despite its immense potential, this technology does come with its associated challenges to data quality, integration, and scalability. As research continues in this area, it becomes very important to build up the robustness and reliability of AI-driven systems so that they are applicable to any kind of aeroponic setup. That is to say, applying artificial intelligence and deep learning in aeroponics is one of the most radical developments in agriculture technology. By utilizing these technologies, farmers could enrich the total efficiency and precision of the aeroponic system for better health, higher yields, and more sustainable cultivation techniques (Sankaranarayanan, 2021). Aeroponics combined with AI has a strong potential for transforming modern agriculture and solving two of the most pressing problems related to global food production as the field keeps evolving.

Significance of AI and Deep Learning in Agriculture

Artificial intelligence and deep learning are making farming more efficient, effective, and resource-saving. It weds advanced algorithms and machine learning models that learn from large datasets which come from a variety of sources including satellite imagery, sensor readings, and climate data. Deep learning is a subdomain of AI that uses neural networks to deduce patterns and predict the outcome from complex data sets. Together, they have transformative potential in agriculture (Narimani et al., 2021).

One of the major strengths of AI and deep learning lies in their capability for crop management optimization. AI-driven tools are able to analyze real-time data related to soil, weather, and plant health to provide relevant actionable insight. For instance, predictive analytics would project risks of disease or pest outbreaks so interventions could be timely, hence eliminating excessive applications of pesticides. Deep learning algorithms can process images of crops, recognizing deficiency symptoms or diseases for focused management practices in precision agriculture, which avoids broad-spectrum applications (Mehra et al., 2018).

Moreover, AI and deep learning guarantee that resources are efficiently managed. It will analyze environmental conditions and crop needs to enable optimization of irrigation schedules and nutrient application timing to avoid wasting water and fertilizers. Such a targeted approach minimizes the costs of operation and supports sustainable farming by reducing the environmental impact. AI is further capable of enhancing productivity through task automation. Planting, harvesting, and monitoring can be carried out with a high level of accuracy and constancy, guided by AI algorithms instructing automated systems. This would compensate for labor shortages and enhance operational efficiency, hence providing better yields and higher-quality produce (Komosa et al., 2020).

Essentially, AI and deep learning are making agriculture smarter, with tools such as decision-making, optimization of resource use, and productivity enhancement. Further advancement can help answer the pressing challenges to food security and sustainability through a more resilient and efficient agricultural sector.

Background: The infusion of artificial intelligence into agriculture, especially through deep learning, is a radical development in the evolution of farming practices. Traditionally, agriculture was based on manual labor and primitive tools to manage farms and resources, and this resulted in inefficiency, undermining yield. Mechanization and automation were followed by modern agriculture that came with technology, but real change comes with the use of artificial intelligence and deep learning.

The ability of AI to process and analyze huge volumes of data therefore opens up new avenues for precision agriculture. Deep learning is a subset of AI exploiting neural networks for pattern recognition and making predictions from complex datasets. This, therefore, enhances farmers' abilities to make sense of soil health, conditions of crops, and environmental factors on which to base their decisions.

It has been catalyzed by inventions in sensor technology, satellite imaging, and data analytics. This would help AI-driven tools to monitor real-time plant health, predict pest outbreaks, and be efficient while conducting irrigation practices and fertilization. Among the rising global demand for food against growing concerns toward the environment, AI and deep learning become solutions to improvements in productivity, resource efficiency, and sustainability in agriculture. This innovation seeks to address labor shortages, wasting of resources, and food security, therefore designing the future of farming.

Scope: AI and deep learning in agriculture discuss their potential with regard to crop management optimization, resource-use efficiency, and process automation. This involves the real-time monitoring and analysis of environmental factors, detection of diseases, and precision agriculture practices that permit more effective targeting of interventions. AI-driven systems extend to automate planting, harvesting, and other agricultural tasks in order to address labor-related challenges. Besides, these technologies support sustainability through improved water and nutrient use efficiency. The application spans several agricultural sectors, from large-scale commercial farming to small-scale farming, with a variety of aims like yield improvement, reduction of environmental impact, and efforts toward global food security.

FUNDAMENTALS OF AEROPONICS

The major principles and advantages associated with aeroponics concern a greatly advanced form of cultivation without soil, where the plants grow in an air or mist medium while their roots are suspended. The basic principles of aeroponics are in the delivery of the key nutrients and oxygen to the plant roots, which are exposed to air and a nutrient-rich mist. There are various advantages in this system over the conventional and traditional practices of cultivation in soils and other alternative methods that do not involve soils(Samikannu et al., 2022).

Principles

i. **Nutrient Delivery:** Plants receive nutrients through a fine mist or aerosol that is sprayed directly onto their roots. This thus controls the composition and concentration of the nutrients.
ii. **Oxygen Availability:** The roots are exposed to air, and therefore there is ample provision of oxygen necessary for healthy growth and fast development of the plants.
iii. **Water Efficiency:** Aeroponics involves very little water compared to traditional methods that use soil. The system re-circulates the excess water hence saving it from wastage.
iv. **Less Disease:** Without soil, there is no disease or pests that are residing in the soil. This accounts for lesser chance of the plants catching diseases and pests, hence less pesticides.

Benefits

i. **Faster Growth:** Receiving more oxygen and better nutrients, the plant growth rate is way faster. Cultivation cycles are shorter.
ii. **Higher Yield:** With an increased growth rate and a better management of the nutrient availability, it results in bigger yields.
iii. **Resource Efficiency:** Less water and fewer nutrients are used with aeroponics than traditional, in-soil methods, which help in reducing operational costs and environmental impact.
iv. **Space Efficiency:** Due to the vertical or compact placement options that aeroponic systems have, they work well within urban agriculture and controlled environments.
v. **Less Labor:** Misting and nutrient delivery can be automated, requiring less manual labor with less maintenance time needed.

The comparison with other soil-free techniques are shown in Table 1.

Table 1. Comparison with other soil-free techniques

Aspect	Aeroponics	Hydroponics	Aquaponics
Growing Medium	No medium; roots suspended in air/mist	Inert medium (e.g., rock wool, perlite)	Water-based with fish and plant integration
Nutrient Delivery	Fine mist directly to roots	Nutrient solution circulates through medium	Nutrients provided by fish waste
Water Usage	Very low, with recycling of mist	Moderate; solution recirculated	Low to moderate; water recirculated
Oxygen Availability	High; roots exposed to air	Moderate; depends on aeration system	Moderate; oxygenation from fish and pumps
Disease Control	Low; minimal risk of soil-borne diseases	Low; fewer issues with pests and diseases	Low to moderate; disease management needed
System Complexity	High; requires precise misting equipment	Moderate; simpler setup	High; involves fish care and balanced ecosystem
Space Efficiency	High; vertical and compact setups possible	Moderate; space needed for medium	Moderate; requires space for fish and plants

Current Applications

Aeroponics is being used in diverse sectors to meet the specific needs of agriculture and horticulture (P. Kumar et al., 2023):

- **Urban Farming:** There is a significant role that Aeroponics plays in the urbanized areas where space is minimal. Vegetables and Herbs are much grown in the urban areas using vertical aeroponic systems like that in the rooftop and indoor farms.
- **Research and Development:** The controlled environment allows researchers to study in detail about the growth, nutrient requirements, and resistance to diseases of plants in this controlled environment. The studies have greatly contributed to the development of different varieties and cultivation techniques of crops.

- **Commercial Agriculture:** Aeroponics is taken up by most of the commercial growers to produce the best quality crops, inputting fewer resources into the crops. It works well specifically to grow high-value crops such as lettuce, herbs, and microgreens.
- **Educational Institutions:** Schools and universities employ the aeroponics system in quite a large way to represent an educational update towards modern technology in agriculture and a view towards sustainability practices.
- **Space Agriculture:** Keeping in mind the efficiency and best-fit nature of the technique for controlled environments, but on the other hand, due to it being highly intensive in terms of space and resources, NASA and other space agencies are researching its application within aeroponic systems in all possible space habitats to grow food.

Hence, aeroponics is a technology that provides real promise in agriculture, offers better growth conditions, spurs efficiency of resources, and is versatile in various applications.

ARTIFICIAL INTELLIGENCE IN AEROPONICS

Role of AI in Modern Farming

Artificial intelligence makes farming more precise, efficient, and sustainable. In aeroponics specifically, its critical role is optimizing the cultivation environment and plant health apart from mechanization. By integrating AI with aeroponic systems, farmers would be able to take better care of the growing condition, consume fewer resources, and raise overall productivity(Boopathi et al., 2023). Some of the key applications where AI enhances aeroponics are:

Figure 1. Role of AI in modern farming

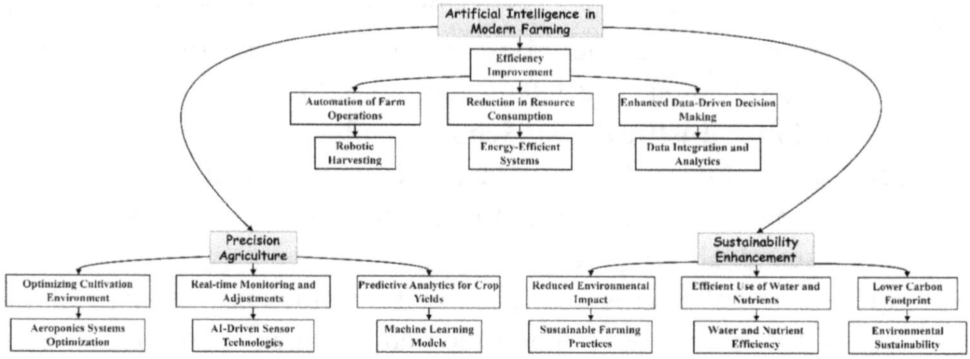

AI is revolutionizing modern farming by improving precision, efficiency, and sustainability, particularly in aeroponics, through real-time monitoring, predictive analytics, and optimization of cultivation environments as shown in Figure 1.

- **Real-Time Monitoring:** AI-powered systems can continuously monitor temperature, humidity, light intensity, and nutrient levels in the environment. Real-time data allows for instant adjustments to be made to meet growing conditions that are optimal.
- **Predictive Analytics:** AI models can project future needs of the plants from historic data and current conditions. It includes projecting the requirement of nutrients, outbreaks of diseases, and the best time for harvest.
- **Automated Control:** AI systems can automate misting, lighting, and nutrient delivery to plants for uniform care, while minimizing the necessity of human intervention. This kind of automation can enhance efficiency while reducing human error.
- **Improved Decision-Making:** AI-driven insight provides actionable recommendations to optimize plant growth and resource use so that farmers can take informed decisions about enhancing yield and quality.

AI-Driven Sensor Technologies

AI-driven sensors will act as a cornerstone to drive the successful application of aeroponics, for they collect various data on different parameters of the environment and plant well-being and relay the collected data. These sensors transmit information on environmental and plant health parameters, which will be analyzed by AI algorithms, affecting changes in real-time.Environmental sensors that are AI-enabled monitor temperature, humidity, amount of CO_2, and light intensity(Naveeenkumar et al., 2024). The sensors will make sure the aeroponics system gives the best growing conditions for the plants; for example, if one of the sensors notes a temperature drop, it automatically engages heating elements.

- **Nutrients Sensors:** It incorporates sensors for mist nutrient concentration that gives the vital elements like nitrogen, phosphorus, and potassium. AI systems work toward adjusting the real-time delivery of nutrients for plants in the best manner possible.
- **Imaging Sensors:** Imaging sensors take high-resolution images of plants to detail the presence of stress, disease, or nutrient deficiencies. Deep-learning-based algorithms—a subset of artificial intelligence—utilize these images to pinpoint any problem at an early age to guide corrective measures.
- **Soil and Root Sensors:** Although no soil is used in aeroponics, there are monitoring sensors that have important parameters of root health and the mist environment. Such sensors should monitor the growth of the roots and mist moisture to give insight into how the plant has developed and how the system is performing.

Machine Learning Algorithms for Data Analysis

Machine learning algorithms are very important in the analysis of the huge bulk of data produced by AI-driven sensors. These algorithms process and interpret data to give valuable insight into aeroponic systems(Sreedhar et al., 2024; Tirlangi et al., 2024; Verma et al., 2024).

i. **Predictive Modeling:** With machine learning models using historical data and real-time data, it gives a possibility of predicting future conditions and needs. An example would be when the predictive models project the level of nutrients required during the different phases of growth and with regard to environmental conditions. This helps in adjusting in advance the amount of nutrient supply.
ii. **Pattern Recognition:** Deep learning algorithms (CNNs) do very well in recognizing patterns in imaging data. With that, it is able to detect the slightest change in a plant's health—things which may be too minute to be captured by the naked human eye, such as diseases and nutrient deficiencies at their early stages.
iii. **Anomaly Detection:** The machine learning algorithms highlight anomalies in the sensor data, possibly indicative of a fault in the aeroponic system. For instance, sudden changes in unexpected nutrient levels or environment conditions can raise an alert for quick corrective actions. Optimization Algorithms: These algorithms optimize the working of an aeroponic system based on acquired data concerning plant growth and resource utilization. These algorithms can adjust the misting schedule, lighting intensity, and nutrient concentrations to achieve maximum growth while ensuring maximum efficiency.
iv. **Adaptive Learning:** The learning of machine learning systems is perpetual and from the newer data; the predictions and recommendations that come back are progressively better. This way, this adaptive learning will ensure the aeroponic system evolves to undergo changeable conditions and needs of the plants.
v. **Decision Support Systems:** AI and ML majorly contribute to decision support systems, which provide farmers with actionable insights and recommendations. These are data-integrating systems that guide decisions on crop management, resource allocation, and system adjustment.

Applications and Impact

Aeroponics, integrated with AI, has huge potential implications for 21st-century farming: enhanced efficiency through streamlining operations and reducing wastages of resources by ensuring that plants have optimum conditions of growth; improved crop quality through constant real-time monitoring and data-driven insights pertaining to the quality of health of a plant; sustainability, given the role of AI in optimizing resource use, therefore making agricultural production more sustainable and less alarming to the environment(Agrawal, Magulur, et al., 2023; Agrawal, Shashibhushan, et al., 2023).

Scalability: AI technology supports the scaling up of operations through aeroponic systems, making it possible to scale up while adapting to environments. Innovation: Further advancements in AI and machine learning are carried through; hence, further innovations in aeroponics into new techniques and technologies have led to farming practices.

This implies that Artificial Intelligence and Machine Learning in aeroponics are the most phenomenal current development in farming. As it were, with AI-driven sensor technologies and data analysis algorithms, farmers would be able to effectively manage aeroponic systems for greater yields of crops and sustainable practices in agriculture. If AI technology further improves, its potential in aeroponics is bound to increase, bringing many more opportunities for innovation and efficiency in farming.

DEEP LEARNING TECHNIQUES

Deep learning is the area within machine learning that involves artificial neural networks with multiple layers connected for modeling and analyzing complex patterns in data. These multilayered networks are referred to as DNNs and typically become very adept in handling vast amounts of unstructured data, like images, text, or time-series data. Deep learning algorithms are specifically designed so that they will automatically learn and extract features from the raw data itself, reducing the need for explicit feature engineering(Venkateswaran et al., 2024).

Figure 2. Deep learning techniques and their components

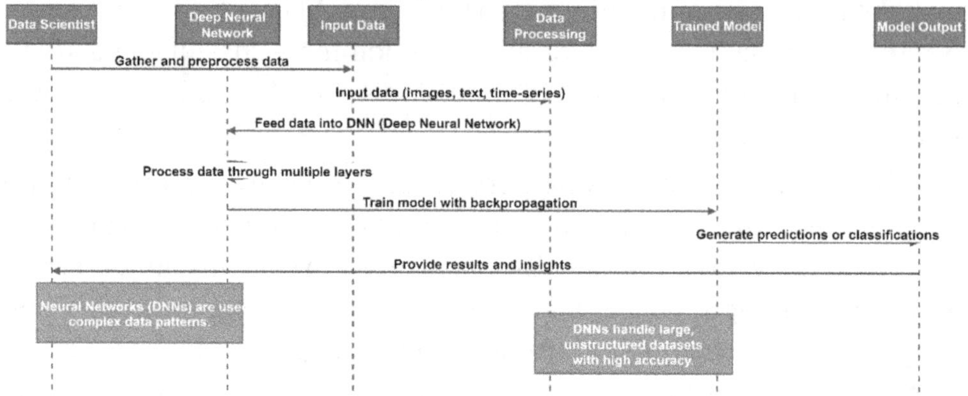

The data scientist collects and preprocesses data, feeds it into a Deep Neural Network (DNN), processes it through layers, and uses backpropagation techniques to adjust weights and biases for accuracy (Figure 2). The trained model generates predictions or classifications based on the input data, and then presents the results for user analysis and interpretation. This workflow illustrates the process of using deep learning techniques in analyzing complex data patterns(Pachiappan et al., 2024).

The foundations of deep learning are primarily based on:

- **Neural Networks:** A neural network includes layers of interconnected nodes (neurons) that process the input through a succession of layers to obtain an output. The layers apply a series of transformations to the data to allow the network to learn hierarchical representations.
- **Activation Functions:** It introduces non-linearity into the model, due to which it is able to learn complex patterns. These include functions like ReLU, Sigmoid, Tanh, which learn whether a neuron should be activated or not with respect to the input data given.
- **Backpropagation:** This is an optimization technique commonly used to update the weights of the neural network according to the error predicted by the network and the actual output. In simple words, it helps the network to learn and try to improve the predictions over time.

CNNs are designed to work on visual and spatial data. Convolutional layers of CNNs automatically detect and learn features from images, hence quite effective for visual tasks. • Recurrent Neural Networks (RNNs): It is designed for sequential data, mostly time-series data and natural languages. It makes use of feedback connections to remember the past, being relevant for applications like language modeling and prediction.

Due to the features of modeling complex relationships and learning from datasets, deep learning has seen a wide base of application in agriculture.

Applications in Plant Growth and Health Monitoring

Deep learning techniques have thoroughly revolutionized monitoring growth and health of plants by providing state-of-the-art plant data analysis, disease detection, and optimum growing conditions (Boopathi, 2024; Glady et al., 2024). Key applications include:

i. **Disease and Pest Detection:** Convolutional neural networks feature in a wide range of plant disease and pest identification. The train is done on large datasets of images of plants with labeled manifestation of various diseases. Such models are then able to detect, with a fair amount of accuracy, the presence of the disease or pest infestation. For instance, CNNs analyze images of leaves to identify fungal infection or insect damage for timely intervention.
ii. **Growth Monitoring:** Deep learning models exploit time-lapse images or video data of plants to monitor the growth patterns and infer the health of the plants. RNNs and LSTMs happen to be key deep models in analyzing sequential data like plant growth over time. Such models would, therefore, predict trajectories of growth and further detect growth pattern anomalies.
iii. **Nutrient and Water Management:** Deep learning techniques process data from sensors that measure the moisture, nutrients, and weather-related environmental variables in the soil to optimize irrigation and fertilization. The neural networks can easily predict the nutrient needs of plants at any stage of growth, according to environmental conditions, and hence provide the right amount of nutrients without wastage.

iv. **Yield Prediction:** Deep learning models predict yields with respect to historical data, weather patterns, and crop characteristics. Deep learning algorithms analyze the patterns in the data and come up with very accurate crop yield forecasts, which will help the farmers decide on harvest timing and resource allocation.

v. **Automated Phenotyping:** Deep learning allows the measurement of and analysis of plant features, which is referred to as automated phenotyping. Through images or 3-D scans of plants, CNN and other deep learning algorithms can estimate leaf area, plant height, biomass, among others—very useful information in breeding and research.

Case Studies and Success Stories

- **Tomato Disease Detection:** CNNs were used by an International Rice Research Institute study for the detection of tomato diseases from leaf images. In this model, the accuracy on identifying many diseases—bacterial and fungal, particularly early blight and late blight, among others—was very high, making the early detection of these diseases easier and thus reducing crop loss. This shows how deep learning can improve plant health management (Boopathi, 2024; P. Kumar et al., 2023; Sankar et al., 2023).
- **Precision Agriculture by IBM:** How Watson can help in precision agriculture has also been explored in terms of crop management optimization. Deep learning merges with weather data, soil conditions, and other crop information to provide actionable insights by Watson. For example, it can predict pest outbreaks, prescribe optimal sowing schedules, and advise on other factors impacting yields and resource efficiency.
- **PlantVillage App:** Penn State University-developed, the deep learning-driven PlantVillage mobile app is for plant disease diagnosis. In essence, the app applies a CNN model trained on thousands of images of plants, thereby letting farmers upload pictures of their plants and receive instant diagnosis and treatment recommendations. The wide reach and acceptance of this app support farmers with real-time information regarding plant health.

- **Crop Yield Prediction Using Deep Learning:** Deep learning algorithms applied by researchers at the University of California, Davis, predicted crop yields with satellite images and climatic data. Their model predicted actual yields so closely that it proved quite effective. Hence, this can be helpful to farmers in making better choices regarding crop management and resource allocation.
- **Deep Learning in Vertical Farming:** AeroFarms, one of the leaders in vertical farming, utilizes deep learning techniques to monitor growth and ensure that it is optimized within its aeroponic system. With data from sensors and cameras, AeroFarms refines environmental conditions and delivery of nutrients, ultimately resulting in more yields and quality produce.

Deep learning technologies have advanced in plant growth and health monitoring, allowing key processes to be automated based on the analysis of large, complex data sets. Deep learning enhances agricultural efficiency and productivity via disease detection, growth monitoring, and yield prediction. Several case studies and successful examples reflect the deep learning-driven transformation underway in these areas with potentiality to solve problems and foster innovation in plant cultivation. The role of deep learning technology is bound to grow with evolution in agriculture, opening up newer prospects for crop management and sustainability.

INTEGRATION OF AI AND AEROPONICS

System Design and Architecture

The integration of AI into aeroponics points at an innovative way to amalgamate sophisticated technology with state-of-the-art agricultural practices. It includes that composite design and architecture to merge AI with an aeroponics system, helping to provide some of the important features to enable collaboration effectively within an intelligent growing environment (Boopathi, 2024; Koshariya, Khatoon, et al., 2023; P. R. Kumar et al., 2023).

Figure 3. Integration of AI and aeroponics, focusing on system design and architecture

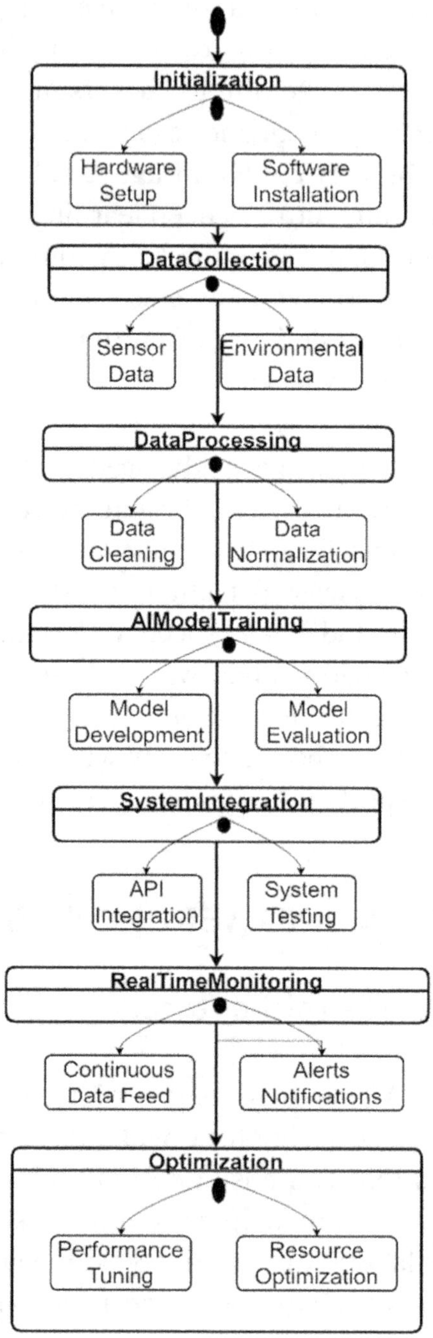

In Figure 3, the initial setup phase involves hardware and software installation, data collection from sensors and environmental sources, data processing, AI model training, system integration, real-time monitoring, and optimization. The process includes data collection, processing, AI model training, system integration, real-time monitoring, and optimization, all aimed at improving system performance and resource usage based on real-time data.

It is made up of a sensor network that monitors various plant growth parameters. It will measure environmental parameters such as temperature, humidity, and light intensity, CO2 and plant-specific metrics like root health, nutrient concentrations, and plant moisture. The sensors will keep on constantly acquiring information and transmit it to the central processing unit.

Data Processing and Storage: The CPU or server will collect all data from various sensors, to be stored using a database or cloud storage system. The architecture should, therefore, be designed to provide high data integrity and availability to fully support real-time access and analysis. There must be developed effective data-storage solutions to handle these huge amounts of data produced by the sensors.
AI and Machine Learning Integration: Algorithms of AI are integrated for the scrutiny of the information collected. Different deep learning models such as Convolutional Neural Network, Recurrent Neural Network, work on processing data for setting particular patterns and prediction for any outcome; whereby, these are made in an online learning mode by being historically trained and incrementally updating with new data to increase its accuracy.
Control Systems: The different aspects of the aeroponic system are controlled under the insights that AI algorithms give: the schedules of misting, the concentrations of nutrients, condition variations of light, and other conditions. Automated control systems ensure optimal plant care with minimum manual intervention.
User Interface and Visualization: This will be a user interface for monitoring the system and being available for farmers or operators to do manual adjustments, if deemed necessary. Embedded in the UI are the alerts, recommendations, and data visualizations served by

the AI, as a result of which the user can make a decision and then act accordingly to serve or rectify a system alert.

Real-Time Data Collection and Analysis

Real-time data collection and analysis are the most crucial operations for optimization of the aeroponic system in order to achieve optimum growth of plants(Koshariya, Kalaiyarasi, et al., 2023; Koshariya, Khatoon, et al., 2023; P. R. Kumar et al., 2023). The steps involved here include:

Continuous Monitoring: Various sensors are built inside the aeroponics system to monitor any environment or plant health parameters that continuously provide real-time information on growing conditions and the plant status.

Data Transmission: Transduction takes place where sensors gather data and then transmit it in real-time to a central processing unit or cloud-based platform. Reliable and secure data transmission protocols are put in place to ascertain the integrity of the data and to access it in a timely manner.

Real-Time Analysis: AI algorithms analyze this incoming data with respect to trend identification, anomaly detection, and assessment of plant health in real time. Machine learning models are applied to recognize patterns and offer instant feedback on how the system stands at that particular moment.

Adaptive Adjustments: The system makes adaptive adjustments based on real-time analysis to further optimize growing conditions. For example, if AI recognizes a depression in the levels of some of the nutrients, it might turn up the volume of the nutrient delivery system automatically to maintain optimal concentrations.

Alert Systems: Inbuilt in this real-time data analysis is an alert system that keeps the user updated with every critical issue or deviation from the best conditions. These could be generated due to large changes in environmental factors, pest detections, or equipment malfunction.

Predictive Modeling and Optimization

Predictive modeling and optimization lie at the heart of AI-driven effectiveness and productivity in aeroponic systems. These techniques make use of historical data and sophisticated algorithms in predicting future conditions and optimizing system performance(Boopathi, 2024; Glady et al., 2024; Venkateswaran et al., 2024).

i. **Predictive Models:** Artificial intelligence-driven predictive models use historical data to make any predictions regarding future plant requirements, growth patterns, and challenges that may probably arise. Predictive models are trained with huge datasets like the trends of environmental conditions, level of nutrients, and plant growth patterns in previous years. Such data analytics enables predictive models to forecast the change well in advance and recommend proactive measures.
ii. **Optimization Algorithms:** Optimization algorithms are run to tune the parameters of the aeroponic system for the best outcome. They do this by varying misting frequency, nutrient concentration, lighting intensity, or other variables in line with predictive insight. The aim is to foster maximum plant growth and yield while maximizing resource efficiency.
iii. **Scenario Analysis:** Predictive modeling will allow scenario analysis for various potential conditions that will affect plant growth. This provides information on the best strategies on managing the resources and optimizing the conditions for maximum growth.
iv. **Continuous Improvement:** As new data arrives and is analyzed, AI systems automatically upgrade their predictive models in a continuous cycle. It's an iterative process that ensures accommodation of, and adjustment to, changes by the system, improving over time with more accurate predictions and better optimization.
v. **Resource Efficiency:** Predictive modeling helps optimize resource use, including water and nutrients. AI reduces wastage and assures that resources are used effectively to ensure sustainability and cost savings by estimating the needs of plants in advance and tuning the delivery systems accordingly.

Rapehay Integration of AI with aeroponics presents a paradigm shift in farming by putting state-of-the-art technology together with the innovative methods of cultivation. The system design and architecture support a robust framework for data collection, processing, and real-time analysis. AI algorithms drive predictive modeling and optimization to enhance plant growth and resource efficiency. On the backbone of real-time insights and adaptive control, AI-powered aeroponic systems enhance productivity, reduce waste, and support sustainable agriculture. Provided technology keeps up with its progress, AI will be, with its future aeroponics integration, a game-changing element in the future of farming and how to solve the global food challenge.

CASE STUDIES

Innovative Aeroponic Farms

AeroFarms

AeroFarms is one of the most popular innovators in the field of aeroponics, which it uses as a part of its vertical farming operations. It has one of the largest indoor vertical farms globally, growing leafy greens and herbs through aeroponics in an extremely controlled growing environment. It executes farming processes with integrated AI and machine learning to achieve farming excellence (KAV et al., 2023; P. Kumar et al., 2023).

AI Implementation and Outcomes: AeroFarms relies on AI-driven sensors and data analytics to track constantly, and through that information, adjust light, temperature, and nutrient levels to create the best possible growing conditions. The machine learning algorithms use data from these sensors for the optimization of the plants' growth conditions and to predict what the plants might require at a given point in time. By dynamically adjusting in real-time, the system has been able to achieve higher crop yields while decreasing resource consumption and operational costs.

The continuous monitoring of data and the requirement for robust AI models that can accommodate change are some of the key lessons learned in this process. Since AI is linked with aeroponics, there has to be a fine line balanced between complete automation and human supervision for a reliable system that produces quality crops. The best practices are investment in high quality sensors and the flexibility of the AI framework itself in accommodating updates with new data.

Plenty

Plenty is also one of the key leaders in aeroponics and conducts vertical farming for producing fresh and top-quality vegetables. It employs a proprietary aeroponic system, together with artificial intelligence technology, to boost productivity and efficiency(KAV et al., 2023).

AI Implementation and Results: Plenty uses AI to control lighting, humidity, delivery of nutrients, etc., for its system. Several plant varieties are given an optimal growth environment by the AI algorithms analyzing data from a number of sensors. This has resulted in a significantly higher crop yield and efficient use of resources than expected by traditional farming methods.

The most important lesson learned was that there has to be an AI infrastructure capable of scaling up with very large data volumes from multiple sensors. It has been critical to use predictive modeling for plant needs and potential problems in optimizing their performance. Integrating AI seamlessly with the operational processes and making adaptations within the system for various plant types and growth stages are some best practices.

AI Implementation and Results

Gotham Greens

Gotham Greens Overview: Gotham Greens grows vegetables hydroponically and aeroponically in city-based greenhouses. Artificial intelligence is at the core of all operations; continuous optimization of plant growth, management of resources. AI Implementation and Results: Gotham Greens handles plant health, provides the necessary nutrition, and regulates the environment through AI-driven analytics. Integrating AI into farming practice has yielded a number of benefits: better crop quality, reduced water and nutrient loss, and improved overall operational efficiency. It is this data-driven approach that allows for growing conditions to be created with such precision that steady, high-quality crops are produced.

Urban Cultivator

Overview: Urban Cultivator designs, manufactures, and sells aeroponics systems for urban applications, such as homes or small commercial businesses. Its systems bring aeroponic farming to all peoples and make it efficient.

AI Implementation and Results: Urban cultivator, in its aeroponic systems, uses artificial intelligence to run light, humidity, and nutrient levels. This AI-driven approach has produced enhanced growth rates and improved overall plant health. It has been very instrumental in ensuring performance consistency and high-quality yields due to its characteristic of responding and adjusting the operations in real-time.

Coupling aeroponics with AI reaps several benefits in crop yield, resource efficiency, and operational efficiency. Such cutting-edge farms as AeroFarms, Plenty, Gotham Greens, and Urban Cultivator are all examples of the power of AI in farming. Case studies from these experiences emphasize key takeaways in the areas of accuracy in data, scalability, user training, and fine-tuning of AI-driven aeroponic systems.

CHALLENGES AND SOLUTIONS

Technical and Practical Challenges

Data Integration and Management: Challenge: Management and integration of huge amounts of data being thrown off by sensors comprise one of the primary technical challenges in the area of AI-integrated aeroponics. It contains environmental variables, plant health indicators, and system performance metrics. Ensuring accuracy, consistency, and timely processing of data can get complex when large datasets are considered. Solution: This would involve the implementation of robust data management systems and cloud-based solutions to handle vast volumes of data. State-of-the-art data integration platforms will facilitate the integration of data from multiple sources, having real-time synchronization that will improve the accuracy of data and make them more accessible. At the same time, data validation techniques and frequent auditing of the system will minimize the errors related to data integrity (Gnanaprakasam et al., 2023; Koshariya, Kalaiyarasi, et al., 2023; Koshariya, Khatoon, et al., 2023; P. R. Kumar et al., 2023).

Sensor Reliability and Calibration: Challenge: Aeroponic systems require accuracy and reliability in their measurements since AI algorithms are informed by them. However, by their very nature, sensors tend to drift over some time, resulting in inaccuracies. This creates a resource-intensive scenario for calibration and maintenance to regularly keep them at optimum performance. Solution: A routine maintenance and calibration schedule would sustain the accuracy of the sensors. Self-calibration technologies or using redundant sensors to cross-check measurement data for reliability can be some of the approaches followed. Another way to enhance system robustness would be by using machine learning models that adapt to small deviations of the sensors.

System Complexity and Integration: Integrating AI into an aeroponic system poses the challenge of increasing system complexity. AI algorithms, sensors, and control systems should communicate very well, which can prove highly challenging. In this regard, compatibility and integration issues may hinder proper functioning. Solution: The use of standardized protocols and modular design in the system, and interoperable technologies and middleware solutions for resolving compatibility issues that can allow easier integration, are recommended. This might reduce integration problems by using comprehensive testing and iterative development for the sake of system coherence.

Cost and Investment: Challenge: Setting up an AI-driven aeroponic system requires substantial up-front investments in technology, infrastructure, and expertise. Advanced sensors and AI software are costly, as is their integration into the system. This has become a major deterrent for most players. Solution: Manage costs with regard to phased implementation strategies and pilot projects where gradual scaling is allowed. Partnership with technology providers and scouting for grants and subsidies from the government may also help in sharing the costs. Moreover, this would justify such an investment if long-term cost savings and productivity gains could be demonstrated.

User Adoption and Training: Challenge: AI-integrated aeroponic systems demand special knowledge and competencies in their operation. This may leave training users to use and service complex systems challenging, more so for those who do not have technical backgrounds. Solution: Comprehensive training programs can help the users easily adopt it. Continuous support and resources need to be given to users in terms of tutorials, help desks, and other such facilities for the proper understanding of the system. It also builds a community of practice and knowledge sharing among users to enhance the development of expertise and confidence.

Strategies for Overcoming Obstacles

Advanced data analytics techniques can develop solutions to the problems associated with data management and integration. Real-time analytics and AI-driven insights provide ease in processing data and give actionable recommendations. Data visualization tools allow the user to extract meanings from complex data to infer a correct inference(Boopathi, 2024; Koshariya, Khatoon, et al., 2023; P. R. Kumar et al., 2023; Pachiappan et al., 2024).

Figure 4. Strategies for overcoming obstacles

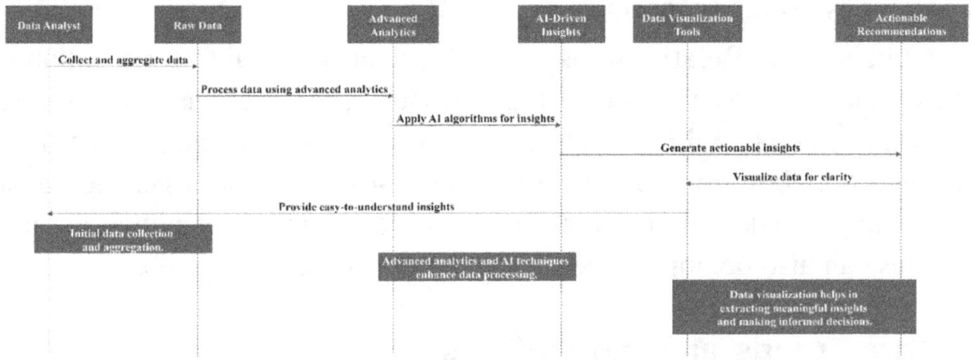

In Figure 4, the user collects and aggregates raw data from various sources, processes it using advanced analytics techniques, and applies AI algorithms for insights. These AI-driven insights interpret the data and uncover patterns. Based on the analysis, actionable recommendations are generated. Data visualization tools present these insights in an easy-to-understand format, providing clear, actionable insights to the user.

Designing Modular Components and Scalable Architectures: The very nature of the system could make the integration easier, whether through modular components or by scalable architectures. This would allow for flexible and extensible technologies, thereby enabling smooth upgrading and tuning. Ensuring the compatibility of the components of a system and well-documented interfaces eases integration, reducing the possibility of problems.

New Models of Funding: From venture capital and crowdfunding to public-private partnerships, there are plenty of alternative models of funding that can be explored in order to help transcend the financial barriers. Return on investment and potential gains, driven by AI in aeroponics, can be highlighted for engaging investors and other relevant stakeholders. Technology adoption can also be encouraged through financial incentives and subsidies.

Continuous Improvement and Feedback: Continuous feedback mechanisms and ways of improving could help in addressing the limitations of the system. AI models update can be regularly done with new data and user feedback to attain better accuracy and reliability. Iterative development practices can aid in refining system functionality and usability through continuous testing by users.

Collaborative Relationships: It can be equally useful to form collaborative relationships with technology providers, research institutions, and industry experts. A lot of help and resources can also be forthcoming through collaboration, which may share knowledge, help get the latest technologies, and jointly work over problems. Industry networks and forums can also be approached for insights and best practices.

Future Trends and Innovations

Advanced AI Algorithms: Future developments in AI algorithms—more sophisticated machine learning models and deep learning techniques—shall support increases in the capability of aeroponic systems. Innovations in AI will provide more accurate predictions, enhanced optimization, and better adaptation to change(Sampath et al., 2022).

IoT and blockchain integration: Internet of Things technologies can be integrated into aeroponics to provide more comprehensive data collection and monitoring. IoT sensors are capable of providing real-time information about the performance of a system and health status of plants. Blockchain technology can bring much-needed security and transparency into the data and transactions.

Automation and Robotics: In most likelihood, not too far in the future, aeroponic systems will depend a lot on automation and robotics. Advanced robotic systems are capable of handling operations related to planting, harvesting, and maintenance, thereby rendering least scope for manual labor. It automates efficiency, minimizes errors, and provides enhanced control over growing conditions.

Sustainable Practices: The quest for sustainability will lead to innovations in aeroponics, which include the design and development of energy-efficient systems, waste reduction technologies, and more sustainable nutrient solutions. Integration with renewable sources of energy, along with resource usage optimization, is going to bend aeroponic farming toward an eco-friendlier, cost-reduced process.

Improved User Interfaces: Probably the most futuristic innovations most likely will be more intuitive and user-friendly interfaces that implement augmented reality and virtual reality. This interface will provide an immersive, interactive experience in monitoring and control of an aeroponic system and make interaction with this technology easier.

This means that the integration of AI and aeroponics would be a matter of challenges as much as opportunities. Embracing future trends and bringing strategic solutions to technical and practical barriers will further enhance the effectiveness and sustainability of AI-driven aeroponic systems. Continuous innovation and adaptation in that sense will be important ways forward to handle challenges and put smart agriculture one step ahead.

FUTURE PROSPECTS IN AEROPONICS AND AI

Future Prospects

AI in aeroponics presents some transformational opportunities for agriculture: enhanced efficiency, sustainability, and productivity(Agrawal, Shashibhushan, et al., 2023; Gnanaprakasam et al., 2023; Koshariya, Kalaiyarasi, et al., 2023). Recently, as technology continues to further advance, there are quite some interesting possibilities regarding the future of aeroponics and AI:

Increased Automation: Without automation, the future of aeroponics would be incomplete. Powered by AI, sophisticated robotic systems are envisioned to perform routine tasks, from planting and pruning to harvesting, much more accurately and efficiently than ever before. This will help reduce labor costs, minimize human error, and improve the overall performance of the system.

Figure 5. Future prospects in aeroponics and AI

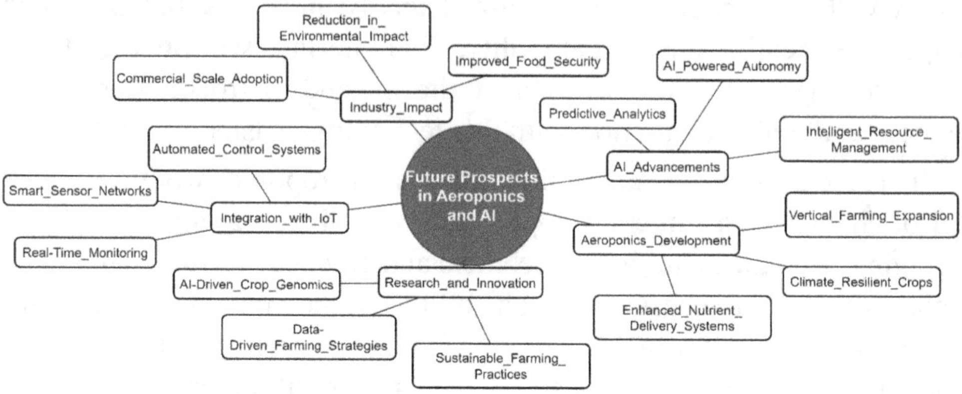

AI-driven precision agriculture: With increasingly accurate predictions and optimizations, AI algorithms will go on refining precision agriculture. Improved machine learning models will analyze vast reams of data to further fine-tune environmental conditions, nutrient delivery, and plant health management. This will increase crop yield and resource efficiency. The figure 5 depicts the future prospects in aeroponics and AI.

IoT and Blockchain Integration: IoT sensors that integrate into aeroponics will allow for the monitoring of a myriad of parameters in real-time and remotely. This adds, through blockchain technology, a layer of security and transparency to obtained data about plant health, nutrient levels, and the performance of the system.

Aeroponics and sustainable energy: This is the future of aeroponics—innovations aimed directly at reducing energy use and minimizing wastes. Advancements in energy-efficient technologies and tapping into renewable energy sources will further make aeroponic systems more environmentally friendly and cost-effective.

Improved User Interfaces: Future technologies like AR and VR will improve user interfaces for practitioners interacting and managing aeroponic systems. These immersive tools provide detailed visualization and simulation capabilities, improving decision making and system management.

Recommendations for Researchers and Practitioners

Invest in Collaborative Research: Researchers and practitioners across disciplines need to collaborate and create innovation in aeroponics and artificial intelligence. Technology developers working in partnership with agricultural experts and academic institutions will foster cutting-edge solutions and knowledge exchange in these specified areas(P. Kumar et al., 2023).

Focus on Data Quality and Integration: AI-driven aeroponic systems are only as good as the data they collect and integrate. Advanced sensing tools and data management investment, therefore, become critical in order to make AI algorithms more accurate and reliable for better performance and its outcomes.

Emphasize Sustainability: Any future research in aeroponics should emphasize sustainability. Inventions of energy-efficient technologies, waste reduction, and integration of renewable energy sources will give way to more sustainable and economically viable systems.

Ease of Use: User-friendly technologies with intuitive interface design will increase the usage and adoption of aeroponics. AR and VR can enable immersive and interactive experiences that allow a better overview for the user in managing and optimizing his system.

Stay Updated With New Emerging Technologies: The professional should be up-to-date about new emerging technologies and trends in AI and aeroponics. Only the ability to continuously learn and adapt to new technology will let one use new developments to one's advantage, ensuring a competitive edge in their respective field.

Aeroponics and AI are even more interesting together, in respect of the future. Innovation in automation, precision agriculture, sustainability, and user experience will drive many major strides forward in crop production and resource management, especially for researchers and practitioners who have taken a keen focus on these areas.

CONCLUSION

The integration of aeroponics with artificial intelligence is one giant step taken in modern agriculture for efficiency, productivity, and sustainability. In this chapter, we have gone through the basics of aeroponics and how AI can help better these systems, also mentioning state-of-the-art techniques in the enhancement of farming.

Aeroponics is a kind of cultivation in which plants are grown in an environment of mist without any soil. It offers a host of advantages over conventional farming, such as better absorption of nutrients, faster growth rates, and fewer resources used. Thus, its efficiency and adaptability make it very relevant for future agricultural production. The integration of AI has huge potential for better management and optimization of the growing conditions in aeroponics. Sensor technologies, machine learning algorithms, and predictive modeling—all AI-driven—make it possible to monitor plant health, manipulate environmental elements, and run automatic processes within the system. Deep learning techniques raise further the AI capabilities within aeroponics to ensure improved prediction and analysis of plant growth and health.

Case studies illustrate how these techniques are used to guarantee successful outcomes for new innovative aeroponic farms. Any technical and practical challenges must be dealt with; therefore, overcoming such issues in this field of aeroponics with the use of AI is critical: data management, sensor reliability, and system integration. Only then will AI in aeroponics be able to realize its full potential. Effective strategies and future trends in this regard have to include enhanced automation and sustainable practices in order to surmount the obstacles and help this sector move forward. Aeroponics, combined with artificial intelligence, foretells increased automation, advanced precision agriculture, and increased sustainability in the near future. IoT, blockchain, and user interface-related innovations will be at the top end to mold the next generation of aeroponic systems.

Hence, already impressed is how huge the junction of aeroponics with AI is able to be in changing agriculture. Advancing technologies present a range of exciting opportunities, from greater crop yields to the optimization of resources and mitigation of many of the challenges in modern farming.

There is great hope that with collaborative approaches, sustainability-focused work, and continued education regarding technological innovation, the integration will be wholly harnessed by researchers and practitioners.

REFERENCES

Agrawal, A. V., Magulur, L. P., Priya, S. G., Kaur, A., Singh, G., & Boopathi, S. (2023). Smart Precision Agriculture Using IoT and WSN. In *Handbook of Research on Data Science and Cybersecurity Innovations in Industry 4.0 Technologies* (pp. 524–541). IGI Global. DOI: 10.4018/978-1-6684-8145-5.ch026

Agrawal, A. V., Shashibhushan, G., Pradeep, S., Padhi, S., Sugumar, D., & Boopathi, S. (2023). Synergizing Artificial Intelligence, 5G, and Cloud Computing for Efficient Energy Conversion Using Agricultural Waste. In *Sustainable Science and Intelligent Technologies for Societal Development* (pp. 475–497). IGI Global.

Boopathi, S. (2024). Sustainable Development Using IoT and AI Techniques for Water Utilization in Agriculture. In *Sustainable Development in AI, Blockchain, and E-Governance Applications* (pp. 204–228). IGI Global. DOI: 10.4018/979-8-3693-1722-8.ch012

Boopathi, S., Kumar, P. K. S., Meena, R. S., Sudhakar, M., & Associates. (2023). Sustainable Developments of Modern Soil-Less Agro-Cultivation Systems: Aquaponic Culture. In *Human Agro-Energy Optimization for Business and Industry* (pp. 69–87). IGI Global.

Fasciolo, B., Awouda, A., Bruno, G., & Lombardi, F. (2023). A smart aeroponic system for sustainable indoor farming. *Procedia CIRP, 116*, 636–641. DOI: 10.1016/j.procir.2023.02.107

Glady, J. B. P., D'Souza, S. M., Priya, A. P., Amuthachenthiru, K., Vikram, G., & Boopathi, S. (2024). A Study on AI-ML-Driven Optimizing Energy Distribution and Sustainable Agriculture for Environmental Conservation. In *Harnessing High-Performance Computing and AI for Environmental Sustainability* (pp. 1–27). IGI Global., DOI: 10.4018/979-8-3693-1794-5.ch001

Gnanaprakasam, C., Vankara, J., Sastry, A. S., Prajval, V., Gireesh, N., & Boopathi, S. (2023). Long-Range and Low-Power Automated Soil Irrigation System Using Internet of Things: An Experimental Study. In *Contemporary Developments in Agricultural Cyber-Physical Systems* (pp. 87–104). IGI Global.

Gowtham, R., & Jebakumar, R. (2023). A Machine Learning Approach for Aeroponic Lettuce Crop Growth Monitoring System. *International Conference on Intelligent Sustainable Systems*, 99–116. DOI: 10.1007/978-981-99-1726-6_9

Pandraju, T. K. S., Boopathi, S., Saravanan, P., Rathan, S. K., & Sathish, T. (2023). Hybrid Deep Learning Technique for Optimal Wind Mill Speed Estimation. *2023 7th International Conference on Electronics, Communication and Aerospace Technology (ICECA)*, 181–186.

Komosa, A., Markiewicz, B., Kleiber, T., Mieloszyk, E., & Mieloch, M. (2020). Yield and nutrient status of greenhouse tomato (Lycopersicon esculentum mill.) grown in new and re-used rockwool, polyurethane, nft and aeroponics. *Journal of Elementology*, *25*(2). Advance online publication. DOI: 10.5601/jelem.2019.24.3.1894

Koshariya, A. K., Kalaiyarasi, D., Jovith, A. A., Sivakami, T., Hasan, D. S., & Boopathi, S. (2023). AI-Enabled IoT and WSN-Integrated Smart Agriculture System. In *Artificial Intelligence Tools and Technologies for Smart Farming and Agriculture Practices* (pp. 200–218). IGI Global. DOI: 10.4018/978-1-6684-8516-3.ch011

Koshariya, A. K., Khatoon, S., Marathe, A. M., Suba, G. M., Baral, D., & Boopathi, S. (2023). Agricultural Waste Management Systems Using Artificial Intelligence Techniques. In *AI-Enabled Social Robotics in Human Care Services* (pp. 236–258). IGI Global. DOI: 10.4018/978-1-6684-8171-4.ch009

Kumar, P., Sampath, B., Kumar, S., Babu, B. H., & Ahalya, N. (2023). Hydroponics, Aeroponics, and Aquaponics Technologies in Modern Agricultural Cultivation. In *IGI:Trends, Paradigms, and Advances in Mechatronics Engineering* (pp. 223–241). IGI Global.

Kumar, P. R., Meenakshi, S., Shalini, S., Devi, S. R., & Boopathi, S. (2023). Soil Quality Prediction in Context Learning Approaches Using Deep Learning and Blockchain for Smart Agriculture. In *Effective AI, Blockchain, and E-Governance Applications for Knowledge Discovery and Management* (pp. 1–26). IGI Global. DOI: 10.4018/978-1-6684-9151-5.ch001

Mehra, M., Saxena, S., Sankaranarayanan, S., Tom, R. J., & Veeramanikandan, M. (2018). IoT based hydroponics system using Deep Neural Networks. *Computers and Electronics in Agriculture*, *155*, 473–486. DOI: 10.1016/j.compag.2018.10.015

Narimani, M., Hajiahmad, A., Moghimi, A., Alimardani, R., Rafiee, S., & Mirzabe, A. H. (2021). Developing an aeroponic smart experimental greenhouse for controlling irrigation and plant disease detection using deep learning and IoT. *2021 ASABE Annual International Virtual Meeting*, 1. DOI: 10.13031/aim.202101252

Naveeenkumar, N., Rallapalli, S., Sasikala, K., Priya, P. V., Husain, J., & Boopathi, S. (2024). Enhancing Consumer Behavior and Experience Through AI-Driven Insights Optimization. In *AI Impacts in Digital Consumer Behavior* (pp. 1–35). IGI Global. DOI: 10.4018/979-8-3693-1918-5.ch001

Pachiappan, K., Anitha, K., Pitchai, R., Sangeetha, S., Satyanarayana, T., & Boopathi, S. (2024). Intelligent Machines, IoT, and AI in Revolutionizing Agriculture for Water Processing. In *Handbook of Research on AI and ML for Intelligent Machines and Systems* (pp. 374–399). IGI Global.

Rajendiran, G., & Rethnaraj, J. (2023). Future of smart farming techniques: Significance of urban vertical farming systems integrated with IoT and Machine Learning. *Open Access Journal of Agricultural Research, 8*(3).

Samikannu, R., Koshariya, A. K., Poornima, E., Ramesh, S., Kumar, A., & Boopathi, S. (2022). Sustainable Development in Modern Aquaponics Cultivation Systems Using IoT Technologies. In *Human Agro-Energy Optimization for Business and Industry* (pp. 105–127). IGI Global.

Sampath, B., Pandian, M., Deepa, D., & Subbiah, R. (2022). Operating parameters prediction of liquefied petroleum gas refrigerator using simulated annealing algorithm. *AIP Conference Proceedings*, *2460*(1), 070003. DOI: 10.1063/5.0095601

Sankar, K. M., Booba, B., & Boopathi, S. (2023). Smart Agriculture Irrigation Monitoring System Using Internet of Things. In *Contemporary Developments in Agricultural Cyber-Physical Systems* (pp. 105–121). IGI Global. DOI: 10.4018/978-1-6684-7879-0.ch006

Sankaranarayanan, S. (2021). Applications of artificial intelligence for smart agriculture. In *AI-Based Services for Smart Cities and Urban Infrastructure* (pp. 277–288). IGI Global. DOI: 10.4018/978-1-7998-5024-3.ch013

Sreedhar, P. S. S., Sujay, V., Rani, M. R., Melita, L., Reshma, S., & Boopathi, S. (2024). Impacts of 5G Machine Learning Techniques on Telemedicine and Social Media Professional Connection in Healthcare. In *Advances in Medical Technologies and Clinical Practice* (pp. 209–234). IGI Global. DOI: 10.4018/979-8-3693-1934-5.ch012

Tirlangi, S., Teotia, S., Padmapriya, G., Senthil Kumar, S., Dhotre, S., & Boopathi, S. (2024). Cloud Computing and Machine Learning in the Green Power Sector: Data Management and Analysis for Sustainable Energy. In *Developments Towards Next Generation Intelligent Systems for Sustainable Development* (pp. 148–179). IGI Global. DOI: 10.4018/979-8-3693-5643-2.ch006

Torres-Tello, J., & Ko, S.-B. (2023). Interpretability of artificial intelligence models that use data fusion to predict yield in aeroponics. *Journal of Ambient Intelligence and Humanized Computing*, *14*(4), 3331–3342. DOI: 10.1007/s12652-021-03470-9

Venkateswaran, N., Kiran Kumar, K., Maheswari, K., Kumar Reddy, R. V., & Boopathi, S. (2024). Optimizing IoT Data Aggregation: Hybrid Firefly-Artificial Bee Colony Algorithm for Enhanced Efficiency in Agriculture. *AGRIS On-Line Papers in Economics and Informatics*, *16*(1), 117–130. DOI: 10.7160/aol.2024.160110

Verma, R., Christiana, M. B. V., Maheswari, M., Srinivasan, V., Patro, P., Dari, S. S., & Boopathi, S. (2024). Intelligent Physarum Solver for Profit Maximization in Oligopolistic Supply Chain Networks. In *AI and Machine Learning Impacts in Intelligent Supply Chain* (pp. 156–179). IGI Global. DOI: 10.4018/979-8-3693-1347-3.ch011

KEY TERMS AND DEFINITIONS

AI: Artificial Intelligence.

AR: Augmented Reality.

CNN: Convolutional Neural Network.

CO: Carbon Monoxide.

CPU: Central Processing Unit.

DNN: Deep Neural Network.

IBM: International Business Machines Corporation.

KAV: Knowledge Augmented Vision.

LSTM: Long Short-Term Memory.

LU: Linear Unit (Linear Algebraic Unit).

ML: Machine Learning.

RNN: Recurrent Neural Network.

UI: User Interface.

VR: Virtual Reality.

Chapter 6
Utilizing Aeroponics Techniques for Improved Farming:
Integration of IoT and Smart Farming Systems With Aeroponics

Hemaltha Jegannathan
SNS College of Technology, India

M. V. Tharunika
SNS College of Technology, India

R. Harinandhini
SNS College of Technology, India

S. Udayamani
SNS College of Technology, India

ABSTRACT

It is the air above the land that will shape farming in the future. Modern methods of agriculture such as Aeroponics, which grows plants in an air or mist environment without the need for soil, are reimagining traditional

DOI: 10.4018/979-8-3693-2320-5.ch006

Agriculture in response to growing global issues like climate change, water shortage, and population expansion. This innovative strategy has the power to completely change the way we grow food, improving Agriculture's efficiency, sustainability, ability to adapt to a world that is changing quickly. Aeroponic system integration with artificial intelligence (AI), IoT (Internet of Things), and smart sensors is transforming farming methods. Precise data gathering and real-time modifications are made possible by these sensors to guarantee ideal plant growth. Aeroponic systems can become more productive by using IoT technology to increase resource efficiency and crop yield while decreasing manual labor. The following are the main IoT sensors found in aeroponic systems:

THE FUTURE OF AGRICULTURE IS IN THE AIR

It is the air above the land that will shape farming in the future. Modern methods of agriculture such as Aeroponics, which grows plants in an air or mist environment without the need for soil, are reimagining traditional Agriculture in response to growing global issues like climate change, water shortage, and population expansion (Benke & Tomkins, 2017). This innovative strategy has the power to completely change the way we grow food, improving Agriculture's efficiency, sustainability, and ability to adapt to a world that is changing quickly.

Aeroponic system integration with artificial intelligence (AI), IoT (Internet of Things), and smart sensors is transforming farming methods (Tzounis et al., 2017). Real-time monitoring of environmental factors, nutrient levels, and plant health is made possible by these technologies. Sensor data can be examined to improve growing conditions, leading to increased productivity and less resource waste. With the least amount of human interaction, AI-driven systems can forecast plant growth patterns, modify nutrient delivery, and even identify early indications of plant illnesses, guaranteeing optimal yield (Farooq et al., 2019).

Aeroponic System Components: The Building Blocks of Innovation

A few essential parts of an aeroponic system combine to guarantee that the plants obtain the proper amount of air, water, and nutrients. By suspending the roots in the air and misting them with nutrient-rich water, these elements aim to provide the ideal environment for plant growth (Lakhiar et al., 2020). An overview of the key elements of an aeroponic system is provided below.

1. **Root Chamber**: The enclosed space where plant roots dangle freely.
2. **Misting System**: High-pressure pumps and nozzles that deliver a fine nutrient spray.
3. **Nutrient Reservoir**: Holds the water and dissolved nutrients.
4. **Environmental Controls**: Regulate temperature, humidity, and light.

Sensor Technologies: The Eyes and Ears of Smart Farming

IoT sensors are essential to an aeroponic system because they automate the monitoring and management of numerous system and environmental parameters (Ferrández-Pastor et al., 2018). Precise data gathering and real-time modifications are made possible by these sensors to guarantee ideal plant growth. Aeroponic systems can become more productive by using IoT technology to increase resource efficiency and crop yield while decreasing manual labor (Tzounis et al., 2017). The following are the main IoT sensors found in aeroponic systems:

- **pH levels**: Ensuring optimal nutrient uptake
- **Electrical Conductivity (EC)**: Measuring nutrient concentration
- **Temperature**: Maintaining ideal growing conditions
- **Humidity**: Preventing mold and optimizing transpiration

Did you know? A single aeroponic system can house over 100 sensors, each feeding data to a central control system every few seconds (Sharma et al., 2020)!

Automation Systems: Precision at Your Fingertips

With IoT, automation in aeroponics reaches new heights (Ray, 2017):

- **Smart Nutrient Dosing**: Automatically adjusts nutrient mix based on plant needs.
- **Adaptive Misting Cycles**: Optimizes water and nutrient use.
- **Climate Control**: Maintains perfect growing conditions 24/7.

Smart Nutrient Dosing in Aeroponics

In aeroponic systems, where plants are cultivated with their roots floating in air and misted with a nutrient-rich solution, intelligent nutrition dosing represents a significant advancement in precision agriculture (Lakhiar et al., 2018). Automated fertilizer dosing systems can further optimize this approach, which is already quite efficient in terms of how much water and space it uses. These systems guarantee that plants get the precise nutrients they require at the appropriate moment, greatly enhancing plant health, growth rates, and yields (Shamshiri et al., 2018). But because plant nutrient requirements vary during their growth cycle, hand dosing is inaccurate and may even be wasteful. Conversely, smart nutrient dosing ensures that plants always receive the ideal ratio of nutrients by using real-time data to automatically modify nutrient levels (Farooq et al., 2019).

Figure 1. Working principle of aeroponics growing system

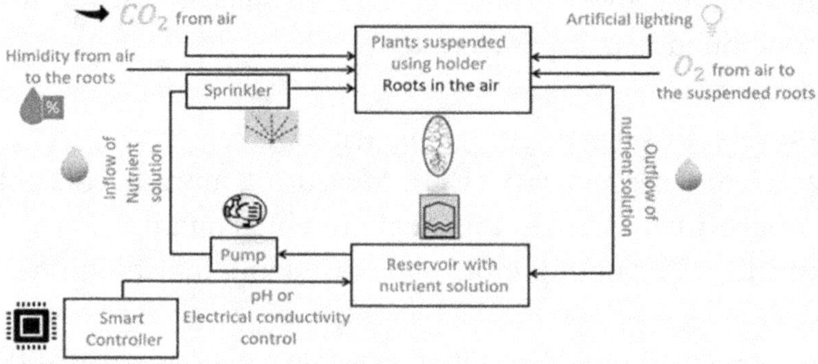

Adaptive Misting Cycles in Aeroponics

In aeroponics, adaptive misting cycles promote plant growth by dynamically varying the length and frequency of misting in response to data obtained in real-time via Internet of Things sensors (Martin et al., 2019). These sensors track variables like temperature, humidity, and root moisture content to figure out the ideal misting intervals. The system adjusts the misting cycles to the plants' changing demands by continuously evaluating this data, making sure the plants get the proper amount of nutrients and water without being over- or under-watered (Shamshiri et al., 2018). This flexibility minimizes water loss, speeds crop development, and enhances nutrient uptake. Aeroponic systems become more productive and sustainable when this leads to improved plant health and more effective use of resources (Elijah et al., 2018).

Climate Control in Aeroponics

In aeroponics, climate management is essential for maintaining optimal environmental conditions that maximize yields and optimize plant growth (Kozai et al., 2020). This entails controlling the growing chamber's humidity, temperature, and CO_2 concentrations. IoT sensors are used by advanced climate control systems to continuously monitor these factors and make necessary adjustments to the ventilation, heating, and cooling systems (Ferrández-Pastor et al., 2018). The ideal moisture levels are maintained by humidifiers and dehumidifiers, while the proper concentration of carbon dioxide for photosynthesis is ensured by CO_2 extractors or injectors. Climate management in aeroponics lowers the danger of disease and pests while promoting consistent plant growth and improving nutrient uptake by establishing a stable and controlled environment (Lakhiar et al., 2020). Because of this accuracy, plants are guaranteed to flourish in any weather, which increases the production and efficiency of aeroponic farming systems.

Crop Yield Improvement: Growing More with Less

IoT-enabled aeroponics isn't just about growing plants—it's about maximizing yield (Ray, 2017):

- **Growth Rate Monitoring**: AI algorithms predict and optimize harvest times.
- **Plant Health Tracking**: Early detection of diseases or nutrient deficiencies.
- **Yield Forecasting**: Data-driven predictions for better resource allocation.

Growth Rate Monitoring

In aeroponics, growth rate monitoring with IoT entails the real-time tracking of plant development through the use of sensors and cameras (Tzounis et al., 2017). Important growth parameters including stem height, leaf area, and general plant health are measured by IoT devices. These sensors send data to a central control system, which analyzes it to look for growth trends and abnormalities in the rates that should be there (Sharma et al., 2020). Based on this information, automated modifications to nutrient distribution, misting schedules, and environmental settings can be made to maximize development. In addition to guaranteeing ideal growth conditions and enabling prompt intervention in the event of problems, this ongoing monitoring also promotes precision agriculture by offering useful insights that improve plant health and production (Elijah et al., 2018).

Plant Health Tracking

Utilizing a variety of sensors and imaging technologies, plant health tracking with IoT keeps an eye on the health of plants grown in aeroponic systems (Martin et al., 2019). IoT sensors track variables like soil moisture, leaf temperature, and chlorophyll content, and AI-enabled cameras identify visual cues of illness, stress, and nutrient shortages (Shamshiri et al., 2018). Real-time insights into plant health are obtained through processing and analysis of the data gathered from these sensors. Growers may be notified of possible problems by automated notifications, allowing for prompt action. IoT solutions assist optimize care, avoid illnesses, and modify growing conditions to improve overall plant performance and production by continuously monitoring plant health (Farooq et al., 2019). This proactive strategy guarantees that plants are regularly observed and kept in their ideal condition, resulting in better crops and more productive agricultural methods.

Figure 2. The aeroponic system using IoT technology by Kerns and Lee (Kerns & Lee, 2017).

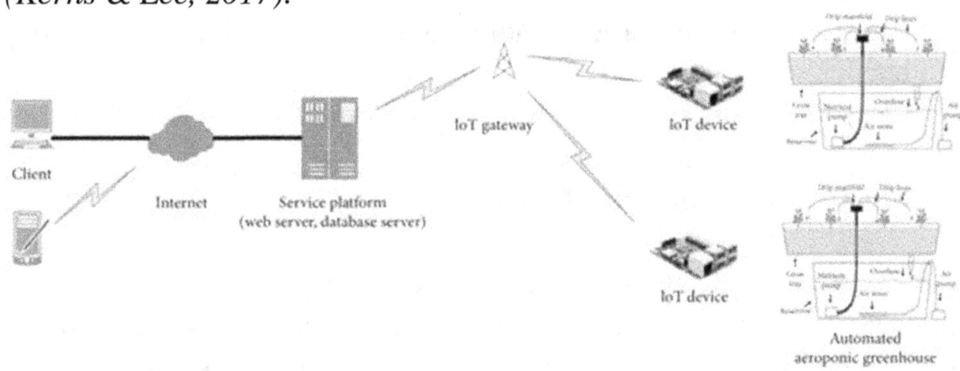

Yield Forecasting

Using real-time data from multiple sensors, IoT-based yield forecasting predicts agricultural yields with a high degree of accuracy (Ray, 2017). IoT devices gather data on variables including past yield history, plant growth rates, nutrient levels, and environmental conditions. This data is processed by machine learning algorithms and advanced analytics to anticipate future yields based on present conditions and trends (Elijah et al., 2018). Farmers can decide on resource allocation, harvest timing, and market strategies by incorporating these forecasts into management systems. This proactive strategy optimizes profitability, reduces waste, and enables input optimization (Tzounis et al., 2017). IoT-based yield forecasting offers insightful information that boosts production and aids in farmers' adaptation to shifting conditions, resulting in more profitable and effective farming operations.

Data Analytics and AI: Farming Smarter, Not Harder

The true power of IoT in aeroponics lies in its ability to turn data into actionable insights (Shamshiri et al., 2018):

Machine Learning Algorithms: Continuously improve growing parameters.
Predictive Maintenance: Anticipate equipment failures before they happen.
Crop Performance Analysis: Identify top-performing plant varieties.

Machine Learning Algorithms

By evaluating the massive volumes of data gathered from sensors included into the growing environment, machine learning algorithms that use the Internet of Things in aeroponics improve system efficiency (Elijah et al., 2018). In order to find trends and connections, these algorithms analyze data on variables including nutrient levels, temperature, humidity, and plant development patterns. Machine learning models can forecast ideal circumstances for plant health and growth by using historical data to learn from past experiences (Tzounis et al., 2017). They can also identify abnormalities and possible problems early on. With the use of this predictive capabilities, climate management, misting cycles, and fertilizer dosage can all be adjusted in real-time to give plants the exact care they require (Farooq et al., 2019). In the end, aeroponics that integrates machine learning with IoT produces better crop yields, more economical use of resources, and a more responsive and adaptive system.

Figure 3. Increase in business operations with machine learning in IoT (Singh et al., 2019).

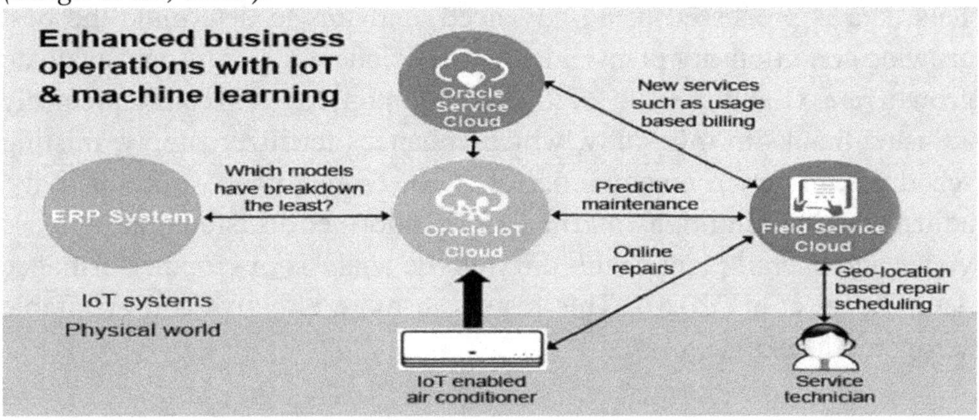

Predictive Maintenance

In aeroponics, predictive maintenance via the Internet of Things (IoT) entails the use of sensors to keep an eye on the health and functionality of vital system parts including pumps, misting nozzles, and climate control units (Sharma et al., 2020). These sensors gather data on temperature, vibration, and operational efficiency on a constant basis. This data is analyzed by machine learning algorithms to find trends that predate equipment malfunctions or inefficiency (Ray, 2017). Predictive maintenance minimizes downtime and avoids expensive failures by anticipating possible problems before they arise. In addition to extending the equipment's lifespan, this preventive measure makes sure the aeroponic system runs well, preserving ideal growing conditions and enhancing plant health and output (Martin et al., 2019). Predictive maintenance, in general, minimizes interruptions while improving system reliability and operating efficiency.

Crop Performance Analysis

In aeroponics, crop performance analysis with the Internet of Things (IoT) integrates sensors and data analytics to continually monitor and assess plant development (Shamshiri et al., 2018). IoT sensors provide real-time data on crop health and development by monitoring parameters including pH, temperature, light intensity, humidity, and nutrient levels. This data is processed using advanced analytics to determine the best growing conditions for plants, identify deficiencies or stress, and evaluate growth rates (Lakhiar et al., 2020). The aeroponic system can be precisely adjusted thanks to this study, which enhances fertilizer supply, misting schedules, and environmental controls. Growers may improve productivity, adjust growing conditions, and make well-informed decisions that increase yields and overall plant health through the analysis of crop performance data (Kozai et al., 2020). This results in more effective and profitable aeroponic farming.

Resource Optimization: Sustainability in Action

Aeroponics is already resource-efficient, but IoT takes it to the next level (Benke & Tomkins, 2017):

- **Water Usage**: Up to 95% less water compared to traditional farming.
- **Energy Management**: Smart systems optimize lighting and pump operations.
- **Nutrient Recycling**: Recapture and reuse nutrients for minimal waste.

Water Usage

Aeroponics employing Internet of Things maximizes water usage by means of automated control systems and real-time monitoring (Lakhiar et al., 2018). IoT sensors provide precise information on water distribution and consumption by monitoring the system's humidity, flow rates, and water levels. By analyzing this data, it is possible to guarantee that plants get the exact amount of water they require, minimizing waste and avoiding over- or under-watering (Elijah et al., 2018). Water efficiency is ensured by automated systems that modify nutrient delivery and misting cycles in response to sensor data. Aeroponic systems can preserve ideal moisture levels while saving water by incorporating IoT technology, resulting in more environmentally friendly farming methods (Martin et al., 2019). This exact management improves plant growth and overall system efficiency while also lowering resource waste.

Energy Management

Using automation and intelligent sensors, energy management in aeroponics with IoT entails optimizing energy use across the entire growth system (Shamshiri et al., 2018). IoT devices track how much energy is used by various components, including misting systems, heating, cooling, and lighting. Machine learning algorithms analyze real-time data on environmental variables and energy use to find patterns and suggest changes to reduce energy waste (Ray, 2017). For instance, the system can improve heating and cooling based on temperature trends or modify lighting intensity based on plant growth stages. This proactive strategy improves system sustainability and efficiency while lowering energy expenditures. Aeroponic operations become more economical and environmentally benign when resources are used effectively, which is ensured by IoT-driven energy management (Kozai et al., 2020).

Nutrient Cycling

Using Internet of Things to accurately control the delivery and recycling of nutrient solutions, nutrient cycling in aeroponics maximizes plant development (Lakhiar et al., 2020). IoT sensors continuously track important variables including pH levels, flow rates, and nutrient concentrations. In order to make sure that the nutrient solution is balanced and modified as necessary to satisfy the plants' shifting needs, this data is examined (Tzounis et al., 2017). With the use of this data, automated systems can efficiently recycle and replenish the nutrient solution, cutting waste and preserving the ideal level of nutrient availability. IoT improves the efficiency of nutrient cycling, encouraging healthier plants, faster growth rates, and larger yields while preserving resources and reducing environmental effect by continually monitoring and modifying nutrient levels (Farooq et al., 2019).

Real-World Success Stories

Vertical Farms: Skyscrapers of Green

In Singapore, a 30-story vertical farm uses IoT-integrated aeroponics to produce over 1,000 kg of vegetables daily, all within the heart of the city (Benke & Tomkins, 2017). This innovative approach demonstrates how urban agriculture can be transformed through the integration of advanced technology and sustainable growing methods.

REFERENCES

Benke, K., & Tomkins, B. (2017). Future food-production systems: Vertical farming and controlled-environment agriculture. Sustainability: Science. *Practice and Policy*, *13*(1), 13–26.

Elijah, O., Rahman, T. A., Orikumhi, I., Leow, C. Y., & Hindia, M. N. (2018). An overview of Internet of Things (IoT) and data analytics in agriculture: Benefits and challenges. *IEEE Internet of Things Journal*, *5*(5), 3758–3773. DOI: 10.1109/JIOT.2018.2844296

Farooq, M. S., Riaz, S., Abid, A., Umer, T., & Zikria, Y. B. (2019). Role of IoT technology in agriculture: A systematic literature review. *Electronics (Basel)*, *8*(6), 701.

Ferrández-Pastor, F. J., García-Chamizo, J. M., Nieto-Hidalgo, M., & Mora-Martínez, J. (2018). Precision agriculture design method using a distributed computing architecture on Internet of Things context. *Sensors (Basel)*, *18*(6), 1731. DOI: 10.3390/s18061731 PMID: 29843386

Kerns, S. C., & Lee, J. L. (2017). Automated aeroponics system using IoT for smart farming. *Scientific Reports*, *7*(1), 3402. PMID: 28611468

Kozai, T., Niu, G., & Takagaki, M. (Eds.). (2020). *Plant factory: An indoor vertical farming system for efficient quality food production*. Academic Press.

Lakhiar, I. A., Gao, J., Syed, T. N., Chandio, F. A., & Buttar, N. A. (2018). Modern plant cultivation technologies in agriculture under controlled environment: A review on aeroponics. *Journal of Plant Interactions*, *13*(1), 338–352. DOI: 10.1080/17429145.2018.1472308

Lakhiar, I. A., Jianmin, G., Syed, T. N., Chandio, F. A., Buttar, N. A., & Qureshi, W. A. (2020). Monitoring and control systems in agriculture using intelligent sensor techniques: A review of the aeroponic system. *Journal of Sensors*, *2020*, 1–18.

Martin, P., Durand-Petiteville, A., Bazile, J., & Moisan, M. (2019). Aeroponics and connected technologies: Sustainable food production systems. *IEEE Access : Practical Innovations, Open Solutions*, *7*, 153662–153671.

Ray, P. P. (2017). Internet of things for smart agriculture: Technologies, practices, and future direction. *Journal of Ambient Intelligence and Smart Environments*, *9*(4), 395–420. DOI: 10.3233/AIS-170440

Shamshiri, R. R., Kalantari, F., Ting, K. C., Thorp, K. R., Hameed, I. A., Weltzien, C., Ahmad, D., & Shad, Z. M. (2018). Advances in greenhouse automation and controlled environment agriculture: A transition to plant factories and urban agriculture. *International Journal of Agricultural and Biological Engineering*, *11*(1), 1–22. DOI: 10.25165/j.ijabe.20181101.3210

Sharma, N., Acharya, S., Kumar, K., Singh, N., & Chaurasia, O. P. (2020). Hydroponics as an advanced technique for vegetable production: An overview. *Journal of Soil and Water Conservation*, *17*(4), 364–371. DOI: 10.5958/2455-7145.2018.00056.5

Singh, V., Sharma, A., & Kumar, M. (2019). Machine learning approaches in IoT agriculture: Current trends and future challenges. *Internet of Things : Engineering Cyber Physical Human Systems*, *8*, 100107.

Tzounis, A., Katsoulas, N., Bartzanas, T., & Kittas, C. (2017). Internet of Things in agriculture, recent advances and future challenges. *Biosystems Engineering*, *164*, 31–48. DOI: 10.1016/j.biosystemseng.2017.09.007

Chapter 7
Integrating IoT and Cloud Computing in Aquaponics and Aeroponics Methods

Pramod Kumar

https://orcid.org/0009-0001-0700-4253

Department of Computer Science and Engineering, Ganga Institute of Technology and Management, Jhajjar, India

Raja Rao P. B. V.

Department of Computer Science and Engineering, Shri Vishnu Engineering College for Women (Autonomous), Bhimavaram, India

A. Ananthi

Department of Computer Science and Engineering, Vel Tech Rangarajan Dr. Sagunthala R&D Institute of Science and Technology, Chennai, India

P. John Augustine

Department of Information Technology, Sri Eshwar College of Engineering, Coimbatore, India

S. Subburam

Department of Information Technology, New Prince Shri Bhavani College of Engineering and Technology, Chennai, India

M. Sudhakar

https://orcid.org/0009-0003-1456-2709

Department of Mechanical Engineering, Sri Sai Ram Engineering College, Chennai, India

DOI: 10.4018/979-8-3693-2320-5.ch007

ABSTRACT

This chapter is just a presentation of the application of IoT and cloud computing to aquaponics and aeroponics, modern technologies used for the sustainable production of food. IoT devices allow real-time monitoring and automation to optimize various conditions that include water quality, nutrient levels, and environmental factors to give crop yields and fish health. Cloud computing allows the storage, handling, and remote assessment of data—hence predictive analytics and informed decision-making. Especially with the help of these technologies, especially systems such as aquaponics and aeroponics, it provides better efficiency, scalability, and resource efficiency especially on being put up against issues like resource scarcity and environmental impacts. The chapter covers technological architecture, implementation strategies, and foreseen benefits, but it also takes into consideration challenges related to data security and system interoperability.

INTRODUCTION

The integration of Internet of Things and Cloud Computing into aquaponics and aeroponics systems represents a landmark development in the field of sustainable agriculture, revolutionizing how food can be produced, monitored, and managed. The continuously increasing world population today needs some innovative ways to produce higher demands of food with less resource application and reduced environmental impacts. In fact, traditional farming involves many constraints, such as the scarcity of land, water, and unpredictable weather conditions. Unlike aquaponics and aeroponics, it will have segments identified as key applications through efficiency controls that are workable, scalable, and not so constrained. The application of IoT in further optimization through individual control, monitoring, and measuring devices for various processes will maximize productivity in the most resource-efficient manner and enable the making of data-backed decisions using cloud computing (Arvind et al., 2020).

Aquaponics is the combination of aquaculture, or raising fish, and hydroponics, or growing plants without soil, in an integrated, symbiotic environment. Nutrients from fish excreta are converted by bacteria in this system into forms that are usable by the plants, purifying water that can be safely reused by the fish. This is a closed-loop, very effective system that saves water and does not need chemical fertilizers. On the other hand, an aquaponics system will only be really successful if the water quality parameters are maintained within a certain level, such as pH, temperature, dissolved oxygen, and nutrients. Now, IoT plays a very important role in providing real-time monitoring and automation. Continuous measurements of the sensors in water and air provide data related to a variety of parameters, which are transmitted to a central system that can leverage this information for automatically regulating the conditions like, for instance, pump control, temperature regulation, and nutrient supply so that the environment is appropriate for the fish and the plants (Karimanzira & Rauschenbach, 2021).

Similarly, aeroponics—growing flora in the air or mist, uncoupled by the use of soil for survival—equally benefits from the application of IoT. In aeroponics, the plants are provided in a closed or semi-closed setting and are misted at some given intervals with nutrient-rich solutions. Due to such a kind of cultivation method, the level of control in the growing conditions allows., therefore, the levels of growth rates and yields to be higher compared to conventional farming methods. However, quite opposite to that is the right misting frequency, nutrient, and environmental factors that play a very vital role in the success of it. The IoT devices that use humidity sensors, nutrient detectors, and temperature monitoring will serve a constant inflow of real-time data through effect automation and optimization processes to provide optimal conditions. The IoT will constantly oversee the environment and set conditions properly so that the plants receive proper nutrients, moisture, and result in a healthy and productive crop (Abbasi et al., 2022).

The cloud computing supporting an aquaponics and aeroponics system, through IoT, empowers more data storage, analysis, and accessibility. In traditional farming situations, data collection and analysis are generally intensive on human resources. In an automated system, data from several sensors is continuously uploaded to the cloud for real-time analysis. Cloud-based analytics tools can find trends, foreseeing the worst failures, and recommending remedial action to the farmer. This allows the farmer to act in a proper way, backed by the insights and forgone assumptions indicated by the data in an easy-to-understand manner. For example, algorithms can reside in the cloud and determine when crops are likely to require increased water or nutrients, based on past crop levels of temperature, humidity, and other nutrient levels, therefore reducing waste and optimizing resource use (Choi et al., 2023).

In addition, Cloud Computing circumspectly permits remote monitoring and management of aquaponics and aeroponics systems. Results may be viewed and the system controlled from anywhere using smartphones, tablets, or computers. This level of remote accessibility becomes very valuable in huge set-ups or when the system is located in a hard-to-reach area. Also, one can collaborate with many other stakeholders, such as agronomists, technicians, and researchers, who can access the same data to facilitate the input of their expertise for the optimization of the overall system. Besides, cloud-based platforms can be integrated with other technologies, such as Artificial Intelligence and Machine Learning, to develop advanced predictive analytics and automation. For instance, a number of systems' data could be used by AI algorithms to analyze and pinpoint best practices, predict crop quantities that are likely to be harvested in the next season, and recommend changes to drive higher efficiency and/or sustainability (Styawati & others, 2023).

Some important challenges it can also address for aquaponics and aeroponics systems relate to the integration of IoT and cloud computing into maintaining their reliability and avoiding equipment failure, which could have disastrous effects on both fish and plants. The devices can monitor the health of equipment like pumps and sensors, which might otherwise be an indication of a problem long before a system failure would occur. The cloud-based analytics empowers predictive maintenance algorithms for analyzing data from several systems and identifies common points of failure, suggesting preventive measures. A proactive approach will cut losses incurred due to downtime and will assure continuity in the operation of systems (Chandramenon et al., 2024).

Another challenge is the need to scale, especially now that demand for food production is on the rise. IoT and Cloud Computing can facilitate the scaling of Aquaponics and Aeroponics systems by providing for centralized management of several of them across different locations. In the cloud, it is possible to aggregate data from several systems whereby operators can compare their performance, work out the trends, and consequently optimize the operations of all of the systems. This scalability is very important in satisfying the increasing demand for locally grown and organic produce within urban areas, where both space and resources are usually limited(Channa et al., 2024).

Therefore, the radical advance to be seen in sustainable agriculture is how IoT and cloud computing are now integrated into aquaponics and aeroponics systems. It is then that these technologies support the optimized use of resources and productivity, realize further scalable solutions in food production, and are operated with the help of real-time monitoring, automation, and data-driven decision-making. Quite the opposite, however, in the case of a fast-growing population coupled with increasing environmental challenges, IoT and Cloud Computing in Aquaponics and Aeroponics are considered mainstream devices that are to feed humanity in a sustainable and environment-friendly manner. These technologies not only offer immediate benefits in terms of efficiency and productivity but also pave the way for future innovations in smart agriculture (Jiang & Liu, 2024).

The chapter discusses the integration of the technologies for Internet of Things and Cloud Computing within an aquaponics and an aeroponics system aimed at establishing the potential for these technologies to radically change the face of sustainable agriculture. It will, therefore, address the technological frameworks, strategies for implementation, and benefits with challenges in integrating these technologies. This provides an overview of how IoT and cloud computing can ensure improved efficiency, scalability, and sustainability of these emerging farming methods. Real-world applications, case studies, and future trends can therefore give a good insight into practical aspects of deploying such technologies into different agricultural environments (Jiang & Liu, 2024).

The topic shall be based on the interjection of the surging demand for sustainable food production systems in the world. Traditional agriculture is facing challenges like water scarcity, land degradation, and a decrease in productivity. To be able to feed the growing human population, aquaponics and aeroponics have come forward as promising solutions for efficient and resource-saving alternatives to conventional farming. Aquaponics is simply a fusion of fish farming with hydroponics, whereby water used by these animals is further channeled back to supply water for the plants, and the plant root intake reciprocally cleans the animals' water. Aeroponics can be considered different in the way plants are cultivated with mist or air surroundings; the difference is that in this system, the roots are suspended and periodically replenished with a nutritive solution. These growing systems are not only more water-wise; the growing system can go from a seasonal outdoor to year-round production using a controlled indoor growing environment (Liu & Jiang, 2024).

The success of aquaponics and aeroponics systems, however, depends on the exact monitoring of such environmental parameters as temperature, humidity, and nutrient levels. Currently, through the advancement of both IoT and Cloud Computing technologies, systems can be monitored in real time for these parameters, facilitating automation of the systems' responses and even allowing better decision-making through real-time analysis of both incoming and obtained data. The integration is a remarkable advance in the domain, as it addresses most of the problems that traditional agriculture was found to have, and even opens up easier pathways to more resilient and sustainable food production systems.

Objectives

This chapter will investigate the merging of technologies for the Internet of Things and cloud computing toward aquaponics and aeroponics systems, in an effort to embrace sustainable agriculture. It aims to explore how these technologies can be used to enhance resource management, increase productivity, and scale these farming methods. On the other hand, the chapter attempts to give some thorough analyses of the technology frameworks and implementations taken and to draw attention to the benefits, difficulties, and uses of IoT and Cloud Computing in these systems.

Case studies and future trends presented in this chapter are also intended to offer some practical approaches along with some recommendations that can be actionable in the deployment process. Ultimately, the current chapter aims to contribute to an overarching vision for the potential of IoT and cloud computing in transforming food production, handling, and distribution to address international agricultural challenges.

FUNDAMENTALS OF AQUAPONICS AND AEROPONICS

Aquaponics is a sustainable farming system in which the rearing of aquatic animals, mostly fish, synergistically combines with the production of crops or plants without using soil or hydroponics. Application of aquaponics is primarily rooted in the natural nitrogen cycle. The fish constantly produce waste in the form of ammonia. If the system was like most ordinary culture-based systems, this waste would continue building up until it attains deadly levels and the fish start dying (Narimani et al., 2021). An aquaponics system, however, will convert the ammonia into nitrites and further into nitrates through beneficial bacteria. These nitrates are fundamental plant nutrients. The plants absorb these nitrates, whereby they filter and clean the water, which is then recirculated back into the fish tanks.

The simplest components that would make up an aquaponics system include a fish tank, a biofilter, grow beds, and a sump tank. The fish tank is where the aquatic animals are reared and formulate the initial wastes. The biofilter is important in hosting the nitrifying bacteria that will convert the harmful ammonia excreted by the aquatic animals to nitrates. The plants, which absorb the nutrients from the water, should be placed in the grow beds, which can be filled with growing media like gravel or expanded clay. At the very bottom, there should be a sump tank that collects water from the grow beds and pumps it back to the fish tank to maintain a continuous flow.

Meanwhile, an aquaponics is usually equipped with monitoring systems of pH, temperature, dissolved oxygen, and nutrient levels in water so that conditions are at their best for fish and plants. Fish stocking density has to be balanced against plant growth. Too many fish generate too much waste; too few plants can't absorb enough nutrients, thus disturbing balance.

Principles and Components of Aeroponics Systems

Aeroponics can be defined as a process of growing plants in air or mist, lacking a soil or an aggregate medium. The most important principle in aeroponics is that the plants are given the required nutrition through a mist of finest droplets, giving a potential maximum air exposure to the roots and hence increasing growth rates and crop yields within a short time. This is particularly resource-use efficient, since its demand for water and nutrients is far much lower compared to conventional agriculture (Rajendiran & Rethnaraj, 2024).

Usually, aeroponics has a plant support structure, which will hold the plants and which has a nutrient reservoir close by, a misting system, and an enclosure or chamber. The plant support structure usually supports the plants either by the root system or with roots suspended in the air. The nutrient reservoir is where the water and nutrient solutions are stored, which is often pumped out and sprayed through the misting system. The fine nozzles or atomizers which are part of the misting system disseminates the nutrient-rich solution to the roots. Some of these enclosures or chambers are often sealed in a way that makes them retain moisture holding humidity and also do not allow the mist to disperse very fast.

An aeroponics system should be sprayed at frequent intervals and be timed for the right duration. By nature, the roots need to be moist but not too wet to supply needs without becoming rotten. The sensing and timing requirements for proper misting of nutrient solutions to the plants are usually automated with an aeroponics system. Additionally, other growth factors, for instance, light, temperature, and CO2 levels, can be tightly controlled in an aeroponic environment, helping to determine the health and productivity of the plant.

Comparative Analysis of Aquaponics and Aeroponics

While both aquaponics and aeroponics are very advanced and sharp soilless cultivation techniques with sharp advantages over traditional agriculture, the base on which they work differs considerably, with distinct elements so suited to various applications and environments (Fasciolo et al., 2023).

Nutrient Source and Cycling: One of the most basic differences between aquaponics and aeroponics is related to the source and cycling of nutrients. In aquaponics, the nutrient comes from fish waste, thus requiring fish and plant balance. The cycle of nutrients in aquaponics is inherently natural and self-sustaining; after the initial establishment, very little external input is required. On the other hand, aeroponics utilizes a premixed solution that has to be sprayed directly onto the roots of the plants. As such, it allows very negligible control of nutrient levels but involves the inputs of solution neustonic factors by its regular input and monitoring to maintain the right level of balance.

Water Use and Efficiency: Both systems are more water-efficient than traditional farming, although aeroponics generally uses less water than aquaponics. In aquaponics, the water is cycled through the system continuously, and some of it is lost through plant uptake and evaporation. On the other hand, with aeroponics, a fine mist of water is sprayed onto the roots, very little of which is wasted, and plants are given only what they need. This makes aeroponics particularly suitable for regions where extreme water scarcity occurs.

Although that makes aquaponics systems complex due to the fact that both plants and fish have to be satisfied, in aeroponics there is only a need and a focus on the plants. A delicate balance has to be struck between the two organisms, and any upset means casualty changes in the system. For this reason, an aquaponics system requires more rigorous monitoring and maintenance compared with an aeroponics system. The main disadvantage is that the misting system of aeroponics can be clogged and will require continual maintenance to operate as usual.

Scalability and Application: This technology can be fitted into several scales, from small household systems up to large commercial farming systems. It has been applied in both rural and urban settings, apart from the aspect that space requirements for fish tanks may limit. Aeroponics: It really has many benefits for settings that are urban, since the setup is actually compact and in environments where space is really at a premium. It is also very easily scalable and can be used in vertical operations, which reveals its suitability to maximize production in small spaces.

Environmental Impact and Sustainability: Both systems are friendly to the environment, but the sustainability is really in the inputs and management. Besides, aquaponics is very sustainable, especially due to the responsible fish feed and the little waste it produces. Aeroponics is resource-efficient but requires synthetic nutrients, derived from sources that might have an environmental footprint. Nonetheless, the closed-loop designs of both systems considerably reduce how much they contribute to environmental pollution on a general perspective compared to traditional farming.

Summarizing, while both aquaponics and aeroponics represent the most advanced ways to sustainable agriculture, they have different aims and conditions. On the other hand, aquaponics fits perfectly for people who intend fish farming, combined with plant cultivation in a self-sustainable chain. On the other hand, aeroponics is designed for maximizing the efficiency of plant growth in smaller-space and water-stressed environments.

INTEGRATION OF IOT IN AQUAPONICS AND AEROPONICS

IoT is assimilated into these systems of aquaponics and aeroponics so that the operational work of these sustainable agricultural methods changes direction completely. It offers real-time monitoring, automation, and control for the betterment of efficiency and productivity in these systems. The IoT architecture for both aquaponics and aeroponics is supposed to link a variety of components through a network that would allow smooth data flow and management (Sankaranarayanan, 2021).

Figure 1. Integration of IOT in aquaponics and aeroponics

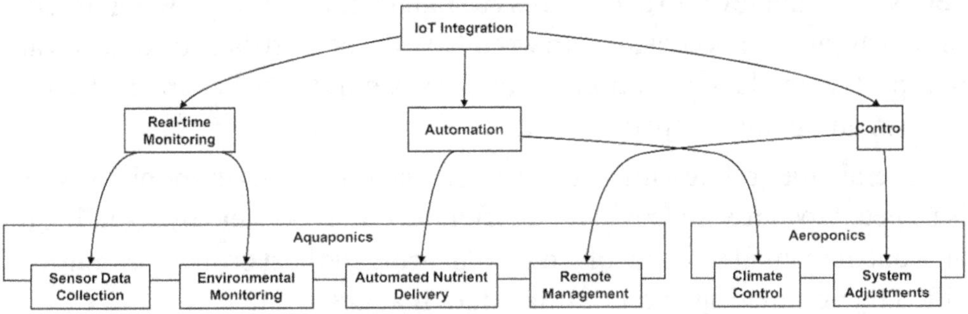

It is, further, the concept of IoT integration into aquaponics, allowing for real-time monitoring, automation, and control, while in aeroponics, as shown in Figure 1, it automates nutrient delivery and climate control. The remote management and adjustments of the system are very useful in both cases.

A typical architecture of an IoT-enabled aquaponics or aeroponics system would encompass sensors, actuators, microcontrollers, communication networks, cloud computing platforms, and user interfaces. Sensors are employed to measure certain environmental parameters, such as temperature, humidity, pH, nutrient concentration, water flow, and light intensity, in such a system. The output sensed through sensing devices is sent on for further processing to microcontrollers, acting as the central processing units by receiving data from sensors and processing the same.

Finally, the processed data is sent to a cloud platform across communication networks: Wi-Fi, Zigbee, or LoRaWAN. The cloud platform is the central location where data is stored, analyzed, and accessed. It facilitates monitoring and control of the system remotely through user end interfaces such as mobile applications or web dashboards. More importantly, the cloud platform can integrate machine learning algorithms with predictive analytics to optimize the performance of a system for actionable insights.

The actuators are an integral component of the IoT architecture and have the prime task of implementing automatic acts in reaction to the obtained data from the sensors. For instance, actuators can switch on heaters or modulate lighting if the temperature inside an aeroponics chamber drops below a certain level to avoid stress to plants during the growing process. Similarly, an aquaponics system would use the sensed data to control water pumps, fish feeders, or nutrient delivery systems to ensure an ideal setting for fish and plants to thrive.

Overall, the architecture of the IoT in aquaponics and aeroponics systems is targeted towards improving the efficiency of the system by reducing the amount of manual labor involved and ensuring the best growing conditions, aided by automation and real-time data management.

Sensors and Actuators for Monitoring and Control

Sensors and actuators are what drive IoT-enabled aquaponics and aeroponics systems to monitor and, in many cases, control various environmental parameters with ultra-high accuracy. These will be very instrumental components in the maintenance of such a delicate balance, which is demanded by optimum plant growth and fish health (Gowtham & Jebakumar, 2023).

Some of the devices put into use in monitoring aquaponics systems include temperature, pH, DO, ammonia, and electrical conductivity sensors. Water temperature sensors ensure that it is within a temperature range that can be stood by the aquatic environment, so as not to stress fish, causing stunted growth. pH sensors monitor water acidity or alkalinity, which is needed by fish and plants. Dissolved oxygen sensors measure the availability of oxygen within the water to provide it in an adequate amount to be breathed by fish. Ammonia sensors monitor ammonia levels that become hazardous due to the excretion by fish. Electrical conductivity sensors measure the concentration of nutrients in water to ensure proper nutrition for plant growth.

It will be equipped with sensors that monitor air temperature, humidity, CO2 level, and pH and conductivity of the nutrient solution. The air temperature and humidity sensors ensure that plants grow in a stable environment since fluctuations will impact growth or development. CO2 sensors track carbon dioxide levels, critical in photosynthesis. The pH and conductivity sensors of the nutrient solution guarantee the proper balance of nutrients the plants require to grow properly.

Some actuators that might be used in aquaponics include water pumps, aerators, fish feeders, and valves. Water pumps drive water from fish tanks to grow beds in a cycle, making sure that the plants are supplied with nutrient-rich water continuously. Aeration equipment increases the level of oxygen dissolved in water, essential in fish respiration. Fish feeders automate this process so that fish get uniform and sufficient feeding. These valves control the flow of water within a system and thus aid in regulating the level of the water and the rate of the flow accurately.

In aeroponics systems, actuators include misting systems, fans, heaters, and lighting controls. Misting systems spray the solution enriched with nutrients onto the plant roots to feed them with nutrition and water. The fans provide air circulation, avoiding excess humidity in the growing compartment for healthy plants. Heaters maintain the proper temperature inside the growing chamber; similarly, lighting controls adjust the light reaching the plants with respect to the daylight cycle.

Real-Time Data Collection and Analysis

Real-time data collection and analysis within IoT-enabled aquaponics and aeroponics systems are very critical for the continuous monitoring and proactive management of the growing environment. Real-time data collection by the system allows for prompt actions against changes in environmental conditions to provide better growth and productivity (Fasciolo et al., 2023).

Data Collection: Sensory devices spread across the aquaponics or aeroponics system continuously collect information on several environmental parameters. This information is then sent to microcontrollers or edge devices, which preprocess this information and then send it to the cloud platform. One can set the frequency by which data shall be acquired—from once every few seconds to several times per hour—based on the system requirements.

Data Analysis: The data, after being sent to the cloud platform, is then reprocessed and analyzed. Very powerful tools that cloud computing offers for data analysis in its service include real-time monitoring dashboards with data visualization coupled with predictive analytics. With these in place, a user is able to monitor trends, identify anomalies, and make informed decisions on system management. For example, in case of a gradual increase in the water temperature, it automatically triggers some cooling mechanism or alerts users.

It can be further applied by machine learning algorithms to continue improving system performance based on the data collected. Conversely, by analyzing historical data, trends of future events with potential implications for plants or fish may be detected—like nutrient deficiencies and temperature changes—and appropriate measures are taken to mitigate the risk of crop failure or fish stress.

Automation and Control: Many processes involved in the aquaponics or aeroponics system can be automatically executed by the IoT system based on real-time data analysis. For instance, if the pH value of an aquaponics system extends beyond the optimal range, then it can automatically add pH-adjusting agents to balance the pH level in the system. Similarly, in the event that the level of nutrients in an aeroponics system decreases, it will automatically be adjusted to the right concentration in terms of nutrient solution so that plants may obtain the proper amount of such nutrients.

Aside from automation, analysis in real-time allows one to monitor and control the system remotely. The user can log into the data and thus control the system from anywhere on a smartphone or computer. This remote access might be quite important either for large-scale operations or when users are running several systems in different locations.

Last but not least, IoT in aquaponics and aeroponics has high potential to increase efficiency, productivity, and sustainability. High precision in the monitoring and control of environmental conditions can be achieved with IoT architecture having state-of-the-art sensors, actuators, and real-time data analytics that optimize growth and minimize the need for human intervention. In fact, these are technologies that will be pushed farther toward revolutionizing sustainable agriculture into accessible, scalable, and efficient practice.

CLOUD COMPUTING IN AQUAPONICS AND AEROPONICS

Cloud computing plays an important role in promoting these aquaponics and aeroponics methods by providing effective data storage and management solutions. Huge amounts of data are normally recorded by a number of sensors in both agricultural methods for environmental conditions, nutrient levels, water quality, and the health of plants. Storage of such voluminous data locally may become daunting due to limited storage and data redundancy. Cloud computing has scaled up this feature by providing secure, cost-effective storage solutions (Venkateswaran et al., 2024).

Figure 2. Cloud computing in aquaponics and aeroponics

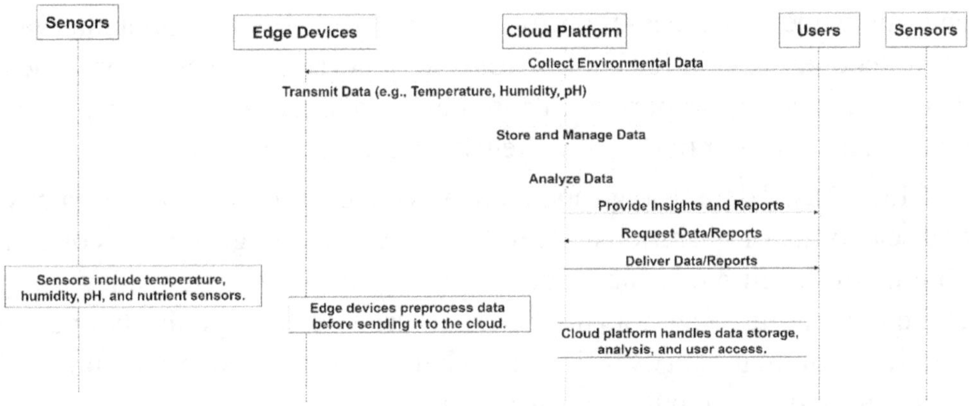

The user uses IoT sensors to monitor any changes in the various parameters of an aquaponics or aeroponics system. In the process, as shown in Figure 2, the data is sent in real-time to a cloud computing platform for analysis. Afterwards, the data is stored and managed by the platform in trends and anomalies. After that, insights and reports are sent back to the user for the adjustment of the system settings. The adjusted data is sent back for continual monitoring.

In a cloud-based system, data from the sensors of both the aquaponics and aeroponics set-ups will be forwarded to remotely hosted servers provided by cloud service providers such as AWS, Google Cloud, or Microsoft Azure. These platforms offer nearly unlimited storage capacity that allows users to store data spanning years without being concerned about running out of space. Moreover, data duplication in the case of cloud storage involves the making of a replica across a number of servers and locations, bringing down the risk of data loss due to hardware failures or disasters to almost nil.

Cloud-based data management allows for effectiveness in organizing and retrieving data. The user organizes and indexes data by time, location, or other measurement types for easy access to subsets of data sets for further analysis. Secondly, the cloud platforms are usually fitted with embedded data management tools that help clean, filter, and aggregate data, hence making it very easy to set up data for analysis.

Another critical aspect of cloud storage is security. Most cloud service providers have a lot of strong security measures in place, which involve encryption methods, access controls, and security audits at regular intervals to protect sensitive information from unauthorized access or breaches. This becomes very important in commercial aquaponics and aeroponics, where data confidentiality and integrity are very paramount.

Cloud-based data storage and management ensure flexibility, security, and scalability of operations that help deal with the huge amounts of data coming out from modern aquaponics and aeroponics systems. By using cloud computing, operators are better assured of the availability and accessibility of their data over time, thus being able to make more informed decisions and continually improve their systems.

Cloud Computing for Remote Monitoring and Control

Cloud computing in aquaponics and aeroponics primarily involves data storage but extends to support remote monitoring and control. In-cloud platforms offer a chance for the user to monitor and control his or her systems from any part of the world where there is some form of internet connectivity, hence flexibility and convenience (Pachiappan et al., 2024).

In a cloud-enabled setup, data from the sensors is transmitted nonstop to the cloud, after which it is processed to show on user dashboards in real-time. The dashboards are accessed over a web browser and often through mobile apps, which allow remote readings for variables, such as temperature, humidity, pH, and nutrient concentration. Cloud computing is done in real time, so a user constantly gets an update of how her or his system works and can in response to altering situations or problems arising.

Another key benefit that cloud computing brings into aquaponics and aeroponics is remote control functionality. With the use of these cloud-based platforms, the user can easily remotely change settings to the system; for example, switching on or off pumps, heating, or lighting depending on data to be received from the sensors. For example, if the user realizes that the temperature in the aeroponics chamber is dropping below a set level, they can turn on the heating system remotely until the perfect growing environment is restored.

Cloud computing offers automation capabilities, which further optimize the efficiency of remote monitoring and control. Cloud platforms can be programmed to trigger a specific action at specific conditions. In this case, at which the pH level in an aquaponics system reaches very low acid levels, the cloud platform is triggered to respond with a pH buffer release to bring the water to a more neutral state, without manual intervention.

It enables the integration of various systems spread over different locations under one unified platform. For instance, operators with large-scale or multi-site aquaponics and aeroponics operations can monitor and control all their systems from one location, reducing the need for personnel on-site and operational costs.

Generally, cloud computing has tremendously improved the extent of monitoring and control remotely affiliated with either the aquaponics or aeroponics operations. Cloud use allows one to hold the perfect conditions where growing is improved, better the response of the system, and reduces the extent of manual intervention at any level, making the operation more effective and productive.

Big Data Analytics and Predictive Modelling

Two of the most transformational applications of cloud computing have turned out to be big data analytics and predictive modeling in aquaponics and aeroponics systems—a way of deriving insightful information from IoT sensors by churning out millions and occasionally billions of lines of data. This fact allows informed decision-making and proactive management of agricultural operations (Sankar et al., 2023).

Big Data Analytics: The sustainability of aquaponic and aeroponic production systems requires real-time information with respect to environmental parameters, plant growth, water quality, and performance of the systems. The volumes of data generated are cumbersomely large and heterogeneous, making it unrealistic to manually handle and process the data. Cloud-based big data analytics tools can process large amounts of data and analyze it in real time, thereby extracting hidden patterns and trends in the data that might not be otherwise apparent.

An instance is big data analytics for determining the link between environmental conditions and plant health, which allows an operator to fine-tune growing conditions using historical information. It can also detect aberrations such as sudden changes in the temperature trend or abrupt pH changes that indicate the potential for system failure or environmental stressors. Operators can identify such issues at an early stage and correct them before they begin to affect crop yield or quality.

Predictive Modeling: Going further than predictive analytics, this method predicts the probability of future outcomes by employing machine learning models with historical data. For predictive models in aquaponics and aeroponics, behavior can be projected under different scenarios; this would be useful in planning and management.

For instance, predictive modeling will have the potential to allow the estimation of the type of seasonal changes impairing performance for the system, in order to rectify operator practices. The models can also be used for predicting the timing of harvesting the crop based on plant growth, thereby ensuring crops are at their quality best when harvested. For the case of aquaponics, predictive modeling can be used to monitor rates of fish growth and to set feeding schedules that permit growth and simultaneously prevent the occurrence of wastage.

In addition, predictive models may be employed to optimize the use of resources such as water and nutrients, predicting how much and when to apply, based on the available and forecast conditions. This increases not only the efficiency of the system but also cuts down on the operational cost and environmental impact.

Big data analytics and predictive modeling are therefore strong tools in enhancing the management of aquaponics and aeroponics systems. With cloud computing, operators can delve deep into the performance of systems, project future outcomes, and optimize their operations toward efficiency and productivity. Therefore, these technologies are the new frontiers in sustainable agriculture, setting the stage for ever more precise and data-driven approaches to food production.

TECHNOLOGICAL ARCHITECTURE AND IMPLEMENTATION

The design of an IoT and cloud-based system for aquaponics and aeroponics implies an integrated architecture of hardware and software parts, all of which have to guarantee efficient monitoring, control, and data management. It seeks to come up with a system that interconnects sensors, actuators, data storage, and user interface seamlessly to give real-time insights and automation capabilities (P. Kumar et al., 2023; P. R. Kumar et al., 2023). Figure 3 depicts the technological architecture and implementation of various technologies.

Figure 3. Technological architecture and implementation

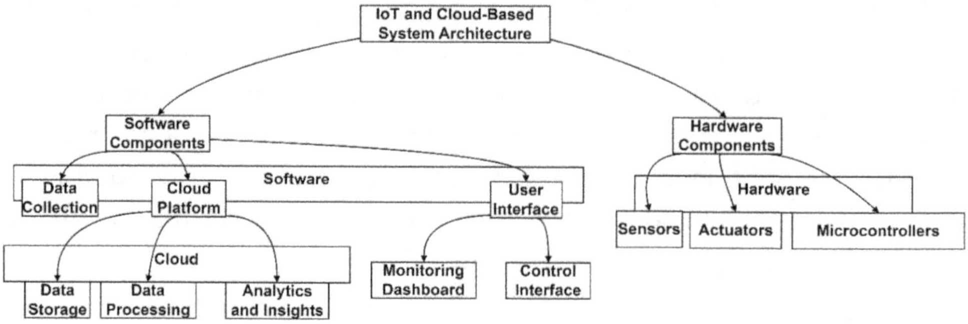

System Architecture: At the very core of our architecture is a network of sensors that pick up data from different points in the aquaponics or aeroponics system. The sensor network would include temperature, humidity, pH levels, dissolved oxygen, nutrient concentration, among other important parameters. These sensors would be connected to microcontrollers or edge devices that process and aggregate data locally before transmission to the cloud.

These microcontrollers serve as an interface between the sensors and the cloud platform by performing pre-treatment tasks, such as filtering, calibration, and formatting of the data. They also handle the communication protocols for the reliable transfer of data to the cloud. This could be microcontrollers such as the Arduino, Raspberry Pi, and ESP32, enabling flexibility and compatibility with several sensors and actuators.

It offers a centralized platform for data storage, processing, and data analysis. This cloud, in which microcontrollers dump their data, actually contains a scalable database. The cloud platform also contains tools in regard to data visualization, real-time monitoring, as well as the control of the system. The way in which a user interacts with this system, then, is via a web or mobile application that will access the cloud platform and allow them to look at data while allowing for alerts and settings to be made within the system from a remote location.

Data Flow and Communication: Data flow in an IoT and cloud-based system is typically multistage. The sensors collect data at all times and pass it to the microcontrollers. The latter preprocess the data and transmit it to the cloud using networks like Wi-Fi, Zigbee, or LoRaWAN. At the cloud level, incoming data are processed with analytics and then stored in databases. Actual uses of the data take place through dashboards and control interfaces hosted at the cloud level.

It will be necessary to support a range of communication protocols and standards to ensure reliable communications. These include MQTT, which stands for Message Queuing Telemetry Transport, providing light-weight messaging between devices and the cloud; and HTTP/HTTPS, including the need to be web-based and, hence, secure. Data encryption and authentication mechanisms are also vital to ensure data integrity and privacy.

Hardware and Software Requirements

Hardware Requirements:

- Sensors: Good quality sensors would be needed to measure environmental parameters such as temperature, humidity, pH, dissolved oxygen, and nutrient concentration. Some of the common used sensors include temperature probes, humidity sensors, pH meters, dissolved oxygen sensors, and electrical conductivity sensors(Koshariya, Khatoon, et al., 2023).
- Microcontrollers, such as Arduino, Raspberry Pi, and ESP32, serve as the central processing units and should therefore be designed to support the sensors that will be included, and later used to handle communication protocols.
- Actuators include water pumps, heaters, lighting, and misting systems for this or that part of the system. These will be selected based on what the particular aquaponics or aeroponics system requires.
- Communication Modules: The communication module will serve for wireless communication: Wi-Fi, Zigbee, or LoRaWAN to send the data from the microcontrollers to the cloud platform. The module to be used will be based on the distance range, data rate, and power consumption. Power Supply: This is going to require appropriate power supply for operating sensors, microcontrollers, and actuators continuously. This can be powered either through batteries or a power adapter.

Software Requirements:

- Cloud Platform: The cloud infrastructure gives data storage, processing, analytics, etc. It should be capable of executing real-time data ingestion and providing data visualization, as well as enabling integration with third-party applications.
- Data Management Tools: Tools to store, aggregate, and analyze data will be necessary, including the storage of historical data—reliable databases and analytics tools to process and visualize data.
- User interfaces: the web and mobile platform applications allow the users to interact with the system, view current data, receive alerts, and control the tools of the system. These interfaces should, therefore, be user-friendly and accessible to all relevant system options.
- It will introduce customized firmware for microcontrollers and software to process and analyze data obtained from the aquaponics or aeroponics system. The microcontrollers are programmed to process the acquired sensor data, conduct cloud communication, and perform operations according to the given control command.

System Integration and Interoperability

This involves putting the IoT and cloud-based segments together in a manner that would ensure interoperability and smooth operations. System integration is essentially about:

- **Component Compatibility:** Each component involved, from sensors to microcontrollers and communication modules, needs to be compatible with all the other components and with the cloud platform. This, in effect, means that sensors will provide data in formats processable by the microcontrollers, while the microcontrollers are able to communicate effectively with the cloud platform using the chose communication protocols.
- **Communication Protocols:** The proper choice of protocols should be made for the communications process to be implemented in a manner such that data transmission takes place appropriately;

hence, system integration will take place. Configuration of protocols like MQTT and HTTP/HTTPS will ensure data is successfully transferred between devices and the cloud. In this way, due to the properly implemented protocols, integrity of the data and system performance will be guaranteed.

- **Standardization of Data:** This is crucial because standardization of the form and unit of data is critical in ensuring that data from the different sensors and components can be compared and analyzed accurately. This step involves the conversion of data into common formats and units before transmission to the cloud for processing.
- **Scalability:** The system design should allow for further expansions or future upgrades, such as increasing the number of sensors, actuators, and monitoring points, and ensuring that the cloud platform scales with increased data volume and processing requirements.
- **Security:** An efficient security system needs to be developed against invasion of the system and leakage of confidential information; this would require encrypting data before transmission, following secure authentication and authorization methods, and keeping the software and firmware updated in order to plug any reported vulnerabilities.
- **System Testing and Validation:** Extensive testing and validation should be carried out to ensure the working of all parts of the system together. This shall involve testing for data acquisition, transmission, processing, and control functionality to troubleshoot any kind of issues before the system's deployment into a live environment.

Such is the case with designing an IoT and cloud-based system for both aquaponics and aeroponics: it brings a system architecture that combines different hardware and software elements. Taken into consideration hardware and software requirements, their interoperability, and robust security arrangements—then it becomes possible to develop a system that will work toward improving the efficiency and effectiveness of aquaponics and aeroponics operations.

BENEFITS OF IOT AND CLOUD COMPUTING INTEGRATION

Inclusion of IoT and cloud computing into these aquaponics and aeroponics systems confers very important benefits; therefore, revolutionizing conventional agriculture practices to improve efficiency, productivity, and sustainability. Such convergence in technology solves some of the pressing contemporary problems with farming and comes up with advanced solutions to drive resource optimization for the attainment of the greater good of food security and conservation of the environment (Agrawal, Magulur, et al., 2023; Agrawal, Shashibhushan, et al., 2023).

Increased Efficiency and Productivity: IoT, coupled with the power of cloud computing, changes the way aquaponics and aeroponics systems work. IoT sensors keep continuous monitoring of a number of environmental parameters of a plant related to temperature, humidity, pH, and concentration of nutrients, providing real-time data to manage them with better precision. The high volume of data acquired by the sensor network is handled and analyzed in the cloud, providing insight into system performance and plant health. Automation tools that include such technologies can adjust the constituent parts—water pumps, heaters, and lighting—to real-time data to ensure optimal growing conditions. This extent of automation, besides saving man-hours, has the automatic effect of reducing human fallibility, hence resulting in increased productivity and more consistent crop yields.

Resource Optimization and Sustainability: IoT and cloud computing advance resource optimization considerably in today's agriculture. These technologies use real-time data to provide control over water and nutrient dispersion, consequently reducing wastage and decreasing the ecological footprint. For example, the IoT sensors in the aquaponics system monitor the level of fish waste and the nutrients released, making it possible to put in just the right quantity of nutrients into the plants without excess usage. In aeroponics systems, misting schedules and nutrient delivery are optimized by cloud-based analytics to conserve water and nutrients. This efficient use of resources will reduce not only operational costs but also support sustainable farming through reduced demand for natural resources and decreased pollution.

Scalability and Flexibility of Systems: The scalability and flexibility the cloud-based architecture presents for the aquaponics and aeroponics systems are hardly matched. In this manner, as the scale of farming operations increases, the data volumes increase, and sensors or components are added; then, they all effortlessly find a home in the cloud platform. Such an attribute guarantees this infrastructure is cost- as well as complexity-aware when the systems scale up. Cloud computing also essentially involves the incorporation of many systems from various locations in order to provide a centralized view of operations with centralized control. This nature supports different scales of operation, be it small-scale urban farming or large commercial setups, based on tailor-made features for each of these setups.

Impact on Food Security and Environmental Conservation. The benefits of Internet of Things and cloud computing do not just acquire the part of operational efficiencies but also extend to social impacts. These technologies increase productivity and consistency in aquaponic and aeroponic systems. Higher efficiency and higher reliability of production methods are important in assuring a continuous supply of fresh and nutritious foods, and this is more than important in urban areas or areas with restricted access to traditional farming. These, along with viable practices encouraged by IoT and cloud computing, help in the conservation of natural resources and avoiding degradation to the environment. All of these technologies do help in ensuring that environmental conservation efforts are adhered to properly and that general ecosystem health is optimal through the use of resources and reduction of waste.

The application of IoT and cloud computing in aquaponics and aeroponics is an instrument through transformation that shall enhance productivity with efficiency and sustainability objectives. In efficiency of resources, it gives seamless scalability, enhancing food security and environmental conversation; they take a front seat in changing modern agriculture to solve global challenges.

CHALLENGES AND CONSIDERATIONS

Although the integration of IoT and cloud computing into these systems comes associated with a number of considerable advantages for both aquaponics and aeroponics, this is also not without challenges or considerations while setting out to implement and successfully operate them. These involve concerns over data security and privacy, technical and operation concerns, cost, and accessibility. Therefore, IoT and cloud computing in aquaponics and aeroponics systems come linked with numerous challenges on data security, compliance, technical, operational, cost, and accessibility issues (P. Kumar et al., 2023; P. R. Kumar et al., 2023; Sankar et al., 2023). Those would include, among others, protection of the data, integration of the system, reliability, maintenance, and accuracy of data, as depicted in Figure 4.

Figure 4. Challenges and considerations of integrating IoT and cloud computing in aquaponics and aeroponics systems

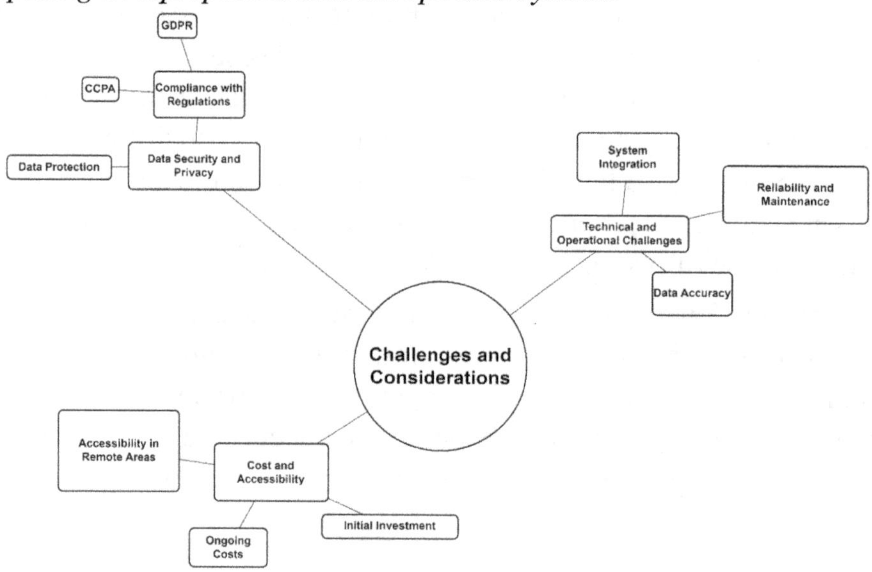

Data Security and Privacy: One of the most prominent issues related to the implementation of both the IoT and cloud computing systems is to ensure that there is a tight security and privacy of the data. The vast amount of data, inclusive of sensitive information on environmental conditions, plant growth, and system performance from IoT sensors, needs to be secured from unauthorized access. The potential targets of cyber attacks include cloud-based platforms that provide convenience and scalability, and the possible cyber threats against them are data breaches and ransomware attacks. It involves making sure that there is encryption of data while in transit or in storage, using strong access controls, and frequent updates of security protocols. Compliance with regulations on the protection of data, such as GDPR or CCPA, therefore becomes very important in securing user data and protecting privacy.

Technical and Operational Challenges: IoT devices and cloud computing present several technical and operational challenges when integrated into the functioning of aquaponics and aeroponics systems. Some of the technical ones are how to make the different hardware elements communicate cohesively or work in tandem and what their communication protocols are, as this can be quite challenging due to the many varieties of sensors, actuators, and cloud platforms. Interoperability between those components is core for uninterrupted data flows and system functionality. Operational challenges include the maintenance and calibration of the IoT devices themselves, which are constantly being monitored for updates to assure accuracy and reliability. Furthermore, system reliability is impacted by problems in network connectivity; as such, it disrupts data transmission and, consequently, system control. Redundant systems and backup solutions can, to a large extent, eliminate the risk of downtime; however, they increase the complexity and cost of a system.

Cost and Accessibility Issues: The cost of implementation for IoT and cloud computing solutions is high and acts as a barrier to entry for small-scale or emerging operations in aquaponics and aeroponics. The cost advantages accrued with purchasing sensors, microcontrollers, and communication modules, followed by cloud services, will be huge. Besides, there are costs for data storage, processing, and system maintenance that have to be continuously incurred. For most users, especially those in developing regions where funds are limited and budgets are always low, the upfront investment in plant equipment, coupled with operational costs, might ultimately prove to be very costly. Similar to this is the technical expertise in setting up, management, and troubleshooting—another limiting factor of accessibility. Therefore, training and support, in addition to developing cost-effective solutions and scalable alternatives, will put this technology within reach of more people.

Leveraging robust security in dealings, proper planning, and viability in costs are central to the attainment of success and hence sustenance in these avant-garde agricultural technologies engendered in the long run.

CASE STUDIES AND APPLICATIONS

Successful application across such a wide variety of settings proves the versatility and impact of the technology. In the following section, we examine a few case studies that bring out the effective utilization of these technologies within urban farms, research institutions, and commercial operations of both aquaponics and aeroponics (Gnanaprakasam et al., 2023; Koshariya, Kalaiyarasi, et al., 2023; Koshariya, Khatoon, et al., 2023).

- **AeroFarms Urban Farm:** Probably one of the most popular companies in the vertical farming business is AeroFarms. It has implemented both the Internet of Things and cloud computing in its aeroponics systems used for farming indoors in urban areas. AeroFarms manages the environment by a highly complex sensor network continuously monitoring temperature, humidity, and levels of nutrition. This gets transmitted back to the cloud for real-time analysis. Such insights allow for the fine-tuning of the aeroponic system with respect to lighting, misting, and nutrient delivery to achieve maximal plant growth while using minimal resources. Second, cloud computing will make it possible for AeroFarms to scale efficiently its operations by easily managing a few farms situated in different places from one platform. This has resulted in increased productivity and reduced operational costs besides increasing sustainability.
- **Controlled Environment Agriculture Center, University of Arizona:** The University of Arizona's Controlled Environment Agriculture Center has utilized both IoT and cloud computing technologies in its research on aquaponics systems. The CEAC applies IoT sensors to measure data regarding water quality, fish health, and

parameters of plant growth. This information gets transmitted into a cloud-based system where researchers do real-time analysis and modeling for the optimization of the aquaponics system. It gives full system performance visibility, thereby enabling researchers to execute many configurations on the cloud platform that improve system efficiency. These integrated technologies deepened the understanding of the dynamics of aquaponics and led to more effective and sustainable farming.

- **Commercial Production:** Growers International is a commercial aquaponics operation that has taken up the IoT and cloud computing challenge for the enhancement of large-scale production capabilities. It shall monitor and control the most critical parameters of the aquaponics system, such as water flow, levels of nutrients, and the environment, using a network of sensors and actuators. These data, collected by the sensors, are subsequently transmitted to a cloud platform where advanced analytics is being applied to optimize system performance and to predict maintenance. By this, Growers International can efficiently manage its production processes, reduce waste, and improve consistency in yield. Cloud computing in this case accommodates remote monitoring and control, thereby facilitating management of various operations from one interface.
- **Singapore Polytechnic:** Singapore Polytechnic has designed its curriculum with an educational initiative that incorporates IoT and cloud computing to provide students with hands-on experience regarding modern agricultural technologies. For instance, in this institution, IoT sensors are utilized in its aquaponics and aeroponics labs in data collection and analysis, system automation, and remote monitoring. This means that data from these systems are then stored in the cloud, analyzed, and students get a chance for hands-on experience in how to make data-driven decisions in running these particular systems. The approach ensures better learning but also prepares them for jobs in smart agriculture.
- **Empowering Smallholder Farmers:** A Sustainable Greenhouse Project in Kenya—Miguel SANE The sustainable greenhouse is one project that supports smallholder farmers in Kenya through the application of IoT and cloud computing. An IoT sensor system

monitors the ecosystem and the prevailing state of crops in a greenhouse. The data flows to the cloud platform for analysis, deriving actionable insights and recommendations for the improvement of crop management. All of these factors are integrated to help farmers optimize resource use, improve yields, and improve the environment. This cloud-based system includes remote access to agricultural experts who offer real-time support to farmers. This application manifests the potential use of IoT and cloud computing in improving agriculture practices in developing regions.

FUTURE DIRECTIONS IN SMART AGRICULTURE

Fast developments in technologies, AI, and machine learning will definitely transform future smart agriculture and change the very approach to aquaponics and aeroponics systems. Such innovations are targeted at creating more efficient, sustainable, and productive agricultural practices and, therefore, stronger and more adaptive ways of farming (Agrawal, Shashibhushan, et al., 2023). The future of smart agriculture will be changed drastically with the advancement in technologies, artificial intelligence, and machine learning, as depicted in Figure 5.

Figure 5. Future directions in smart agriculture

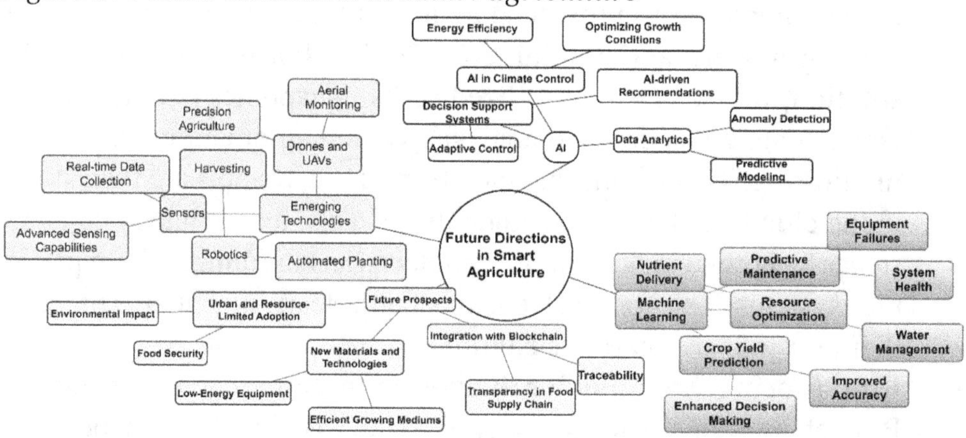

New Emerging Technologies and Innovations: Of the many emerging new technologies, smart agriculture is changing at a very rapid face. Advanced sensors and IoT devices are becoming more sophisticated to allow better monitoring and management of agricultural environments. Indeed, new types of sensors are being developed to provide more precise measurements of soil conditions, plant health, and water quality for better control and more accurate predictions. Moreover, drones and UAVs are increasingly applied in agriculture, turning out to be very effective in air monitoring of crops and aquaponics systems, and their data can be assessed in real time to make proper decisions. At the very heart of this are robotics-related innovations, whereby autonomous systems are created for planting, harvesting, and maintenance of the system itself, further eliminating manual labor and increasing the efficiency of operation.

Role of AI and Machine Learning in Smart Agriculture: Artificial intelligence and machine learning transform smart agriculture through the provision of advanced data analytics and prescriptive capabilities. Running on vast amounts of data from IoT sensors and other sources, AI algorithms are applied to identify patterns and trends, which would otherwise be too difficult for a human to determine with accuracy. These models are used to project the plants' growth and to optimize the resources that are being used. They can also detect problems before they happen or turn into big issues. For instance, AI-powered systems can foretell the need for certain nutrients and deliver them in real-time, thus optimizing the efficiency of such an aquaponics or aeroponics system. More importantly, machine learning algorithms raise the level of precision in control systems for environments, much like the case with greenhouse climate control or nutrient management in aquaponics setups. It allows for better and more proactive decisions in managing an agricultural system, thus increasing yields while reducing resource usage.

In other words, at the moment, emerging technologies, artificial intelligence, and machine learning take the lead in shaping the future of smart agriculture with capability and efficiency overlaps in aquaponics and aeroponics systems. The developments are likely to change the face of agriculture into new chances for sustainable and productive farming. If technology moves further on, such integration will mean considerable developments in agriculture towards resilient, efficient, and environmentally-friendly practices.

CONCLUSION

Hence, the integration of IoT and cloud computing into aquaponics with aeroponics will only usher in a revolution to modern agriculture. This technology will help in achieving new efficiencies, productivity, and sustainability targets. The IoT-based real-time monitoring and control allow for optimization of resource use and system performance, whereas cloud computing leverages scalable data management and advanced analytics. Both enable precision and predictability to drive improvements in crop yield and reductions in environmental impact.

However, they have come at the cost of several security, technical complications, and cost considerations. These challenges have to be straightened out for the full use of the potential carried by the technologies. The future of smart agriculture surely looks promising with the continuous innovation in the area of sensors, AI, and robotics that will help in making the systems of aquaponics and aeroponics even better. These technologies, with the ongoing evolution, will be greatly helpful in countering the global issues of food security and sustainability to produce such results as very resilient and adaptive agricultural practices. Adoption of these innovations is the key to rout the face of agricultural practice to the future for an efficient, sustainable, and productive food system.

REFERENCES

Abbasi, R., Martinez, P., & Ahmad, R. (2022). An ontology model to represent aquaponics 4.0 system's knowledge. *Information Processing in Agriculture*, *9*(4), 514–532. DOI: 10.1016/j.inpa.2021.12.001

Agrawal, A. V., Magulur, L. P., Priya, S. G., Kaur, A., Singh, G., & Boopathi, S. (2023). Smart Precision Agriculture Using IoT and WSN. In *Handbook of Research on Data Science and Cybersecurity Innovations in Industry 4.0 Technologies* (pp. 524–541). IGI Global. DOI: 10.4018/978-1-6684-8145-5.ch026

Agrawal, A. V., Shashibhushan, G., Pradeep, S., Padhi, S., Sugumar, D., & Boopathi, S. (2023). Synergizing Artificial Intelligence, 5G, and Cloud Computing for Efficient Energy Conversion Using Agricultural Waste. In *Sustainable Science and Intelligent Technologies for Societal Development* (pp. 475–497). IGI Global.

Arvind, C., Jyothi, R., Kaushal, K., Girish, G., Saurav, R., & Chetankumar, G. (2020). Edge computing based smart aquaponics monitoring system using deep learning in IoT environment. *2020 IEEE Symposium Series on Computational Intelligence (SSCI)*, 1485–1491. DOI: 10.1109/SSCI47803.2020.9308395

Chandramenon, P., Aggoun, A., & Tchuenbou-Magaia, F. (2024). Smart approaches to Aquaponics 4.0 with focus on water quality- Comprehensive review. *Computers and Electronics in Agriculture*, *225*, 109256. DOI: 10.1016/j.compag.2024.109256

Channa, A. A., Munir, K., Hansen, M., & Tariq, M. F. (2024). Optimisation of Small-Scale Aquaponics Systems Using Artificial Intelligence and the IoT: Current Status, Challenges, and Opportunities. *Encyclopedia*, *4*(1), 313–336. DOI: 10.3390/encyclopedia4010023

Choi, E.-Y., Lee, H.-S., Cha, J. H., & Lee, L.-G. (2023). A Study on the AI Analysis of Crop Area Data in Aquaponics. *The Journal of the Convergence on Culture Technology*, *9*(3), 861–866.

Fasciolo, B., Awouda, A., Bruno, G., & Lombardi, F. (2023). A smart aeroponic system for sustainable indoor farming. *Procedia CIRP, 116*, 636–641. DOI: 10.1016/j.procir.2023.02.107

Gnanaprakasam, C., Vankara, J., Sastry, A. S., Prajval, V., Gireesh, N., & Boopathi, S. (2023). Long-Range and Low-Power Automated Soil Irrigation System Using Internet of Things: An Experimental Study. In *Contemporary Developments in Agricultural Cyber-Physical Systems* (pp. 87–104). IGI Global.

Gowtham, R., & Jebakumar, R. (2023). A Machine Learning Approach for Aeroponic Lettuce Crop Growth Monitoring System. *International Conference on Intelligent Sustainable Systems*, 99–116. DOI: 10.1007/978-981-99-1726-6_9

Jiang, W., & Liu, J. (2024). NeuroAqua: Developing an Optimized Artificial Intelligence and Internet of Things-based Aquaponics System. *2024 IEEE 3rd International Conference on Computing and Machine Intelligence (ICMI)*, 1–5.

Karimanzira, D., & Rauschenbach, T. (2021). An intelligent management system for aquaponics. *Automatisierungstechnik, 69*(4), 345–350. DOI: 10.1515/auto-2020-0036

Koshariya, A. K., Kalaiyarasi, D., Jovith, A. A., Sivakami, T., Hasan, D. S., & Boopathi, S. (2023). AI-Enabled IoT and WSN-Integrated Smart Agriculture System. In *Artificial Intelligence Tools and Technologies for Smart Farming and Agriculture Practices* (pp. 200–218). IGI Global. DOI: 10.4018/978-1-6684-8516-3.ch011

Koshariya, A. K., Khatoon, S., Marathe, A. M., Suba, G. M., Baral, D., & Boopathi, S. (2023). Agricultural Waste Management Systems Using Artificial Intelligence Techniques. In *AI-Enabled Social Robotics in Human Care Services* (pp. 236–258). IGI Global. DOI: 10.4018/978-1-6684-8171-4.ch009

Kumar, P., Sampath, B., Kumar, S., Babu, B. H., & Ahalya, N. (2023). Hydroponics, Aeroponics, and Aquaponics Technologies in Modern Agricultural Cultivation. In *IGI:Trends, Paradigms, and Advances in Mechatronics Engineering* (pp. 223–241). IGI Global.

Kumar, P. R., Meenakshi, S., Shalini, S., Devi, S. R., & Boopathi, S. (2023). Soil Quality Prediction in Context Learning Approaches Using Deep Learning and Blockchain for Smart Agriculture. In *Effective AI, Blockchain, and E-Governance Applications for Knowledge Discovery and Management* (pp. 1–26). IGI Global. DOI: 10.4018/978-1-6684-9151-5.ch001

Liu, J., & Jiang, W. (2024). Optimization of Aquaponics System Efficiency based on Artificial Intelligence Approach. *2024 IEEE 3rd International Conference on Computing and Machine Intelligence (ICMI)*, 1–6.

Narimani, M., Hajiahmad, A., Moghimi, A., Alimardani, R., Rafiee, S., & Mirzabe, A. H. (2021). Developing an aeroponic smart experimental greenhouse for controlling irrigation and plant disease detection using deep learning and IoT. *2021 ASABE Annual International Virtual Meeting*, 1. DOI: 10.13031/aim.202101252

Pachiappan, K., Anitha, K., Pitchai, R., Sangeetha, S., Satyanarayana, T., & Boopathi, S. (2024). Intelligent Machines, IoT, and AI in Revolutionizing Agriculture for Water Processing. In *Handbook of Research on AI and ML for Intelligent Machines and Systems* (pp. 374–399). IGI Global.

Rajendiran, G., & Rethnaraj, J. (2024). Optimizing Lettuce Crop Yield Prediction in an Indoor Aeroponic Vertical Farming System Using IoT-Integrated Machine Learning Regression Models. *Revue d'Intelligence Artificielle*, *38*(3), 825–836. DOI: 10.18280/ria.380309

Sankar, K. M., Booba, B., & Boopathi, S. (2023). Smart Agriculture Irrigation Monitoring System Using Internet of Things. In *Contemporary Developments in Agricultural Cyber-Physical Systems* (pp. 105–121). IGI Global. DOI: 10.4018/978-1-6684-7879-0.ch006

Sankaranarayanan, S. (2021). Applications of artificial intelligence for smart agriculture. In *AI-Based Services for Smart Cities and Urban Infrastructure* (pp. 277–288). IGI Global. DOI: 10.4018/978-1-7998-5024-3.ch013

Styawati, S. (2023). Mustard Plant Growth Monitoring System Using Artificial Intelligence in Aquaponics. *IC-ITECHS*, *4*(1), 37–44.

Venkateswaran, N., Kiran Kumar, K., Maheswari, K., Kumar Reddy, R. V., & Boopathi, S. (2024). Optimizing IoT Data Aggregation: Hybrid Firefly-Artificial Bee Colony Algorithm for Enhanced Efficiency in Agriculture. *AGRIS On-Line Papers in Economics and Informatics*, *16*(1), 117–130. DOI: 10.7160/aol.2024.160110

KEY TERMS AND DEFINITIONS

AI: Artificial Intelligence.

AWS: Amazon Web Services.

CCPA: California Consumer Privacy Act.

CEAC: Controlled Environment Agriculture Center.

CO: Carbon Monoxide.

ESP: Extra Sensory Perception (Note: In tech contexts, it often refers to Espressif Systems, a company specializing in microcontroller units for IoT).

GDPR: General Data Protection Regulation.

HTTP: Hypertext Transfer Protocol.

HTTPS: Hypertext Transfer Protocol Secure.

MQTT: Message Queuing Telemetry Transport.

UAV: Unmanned Aerial Vehicle.

WAN: Wide Area Network.

Chapter 8
Study on Advanced IoT Solutions for Enhancing Agricultural Productivity

Ajay N. Upadhyaya
https://orcid.org/0000-0002-7583-6430

Department of Computer Engineering, SAL Engineering and Technical Institute, Ahmedabad, India

C. Padmaja
https://orcid.org/0000-0003-0521-916X

Department of ECE, G. Narayanamma Institute of Technology and Science, Hyderabad, India

Deepika Dhamija
https://orcid.org/0000-0003-1836-0242

Centre for Distance and Online Education, Manipal University Jaipur, India

Ankit Dhamija
https://orcid.org/0000-0003-4456-9680

Centre for Distance and Online Education, Manipal University Jaipur, India

S. M. Prakash
https://orcid.org/0009-0008-5822-1056

DOI: 10.4018/979-8-3693-2320-5.ch008

Department of Chemistry, St. Joseph's College of Engineering, Chennai, India

ABSTRACT

This chapter explores advanced IoT solutions designed to boost agricultural productivity by integrating cutting-edge technologies. It examines the deployment of IoT sensors and devices for real-time monitoring of soil conditions, crop health, and environmental factors. Key innovations discussed include precision irrigation systems, automated climate control, and predictive analytics that leverage big data and machine learning to optimize crop yields. The chapter also highlights case studies demonstrating successful IoT applications in various agricultural settings, emphasizing the impact on resource efficiency and yield improvement. Challenges such as data security, integration with existing systems, and cost implications are addressed, along with strategies for overcoming these obstacles. By providing a comprehensive overview of current IoT technologies and their practical applications, this chapter offers valuable insights for researchers, practitioners, and policymakers aiming to enhance agricultural productivity through technological advancement.

INTRODUCTION

The agricultural sector is undergoing a profound transformation driven by advanced Internet of Things (IoT) solutions. As global food demand surges due to a growing population and changing consumption patterns, traditional farming practices face significant challenges in terms of efficiency, sustainability, and productivity. The integration of IoT technologies offers a promising solution to these challenges, revolutionizing the way agricultural operations are managed and optimized. By leveraging a network of interconnected devices and sensors, IoT enables real-time data collection, analysis, and actionable insights, which are crucial for enhancing agricultural productivity (Mishra & Sharma, 2023).

One of the primary ways IoT is enhancing agricultural productivity is through precision agriculture. This approach utilizes IoT sensors and devices to monitor and manage field variability in crops. Sensors placed in the soil measure various parameters such as moisture levels, temperature, and nutrient content. This data is then analyzed to make precise recommendations for irrigation, fertilization, and pest control. For instance, smart irrigation systems use real-time weather data and soil moisture levels to optimize water usage, reducing waste and ensuring that crops receive the right amount of water. Similarly, precision fertilization techniques use sensor data to apply nutrients more accurately, minimizing runoff and environmental impact while maximizing crop yield (Dhanaraju et al., 2022).

IoT solutions also facilitate enhanced crop monitoring and management. Drones equipped with high-resolution cameras and sensors provide aerial views of fields, enabling farmers to assess crop health and detect issues such as diseases, pests, or nutrient deficiencies early. This early detection allows for timely interventions, reducing crop loss and improving overall productivity. Additionally, IoT-enabled smart farming equipment, such as autonomous tractors and harvesters, can perform tasks with greater precision and efficiency compared to traditional machinery. These machines are equipped with GPS and sensor technology that allows them to navigate fields autonomously, optimizing planting, tilling, and harvesting processes based on real-time data (Morchid et al., 2023).

Livestock management is another area where IoT solutions are making a significant impact. Wearable sensors and smart tags for animals provide valuable data on their health, behavior, and location. These sensors can monitor vital signs, detect early signs of illness, and track movement patterns. For example, sensors that monitor rumination and feeding behavior in dairy cows can help farmers detect health issues before they become serious, ensuring better animal welfare and increased milk production. Additionally, GPS tracking enables better management of grazing patterns, ensuring that livestock are optimally distributed across pastures, which improves soil health and forage quality (Ponnusamy & Natarajan, 2021).

The integration of IoT in agriculture also supports better resource management and sustainability. Smart irrigation systems, for example, help conserve water by applying it only when and where it is needed, reducing wastage and preserving this critical resource. Similarly, IoT solutions for monitoring and managing energy use in greenhouses and livestock facilities can lead to significant reductions in energy consumption and greenhouse gas emissions. By providing detailed insights into resource usage, IoT technologies enable farmers to make informed decisions that promote sustainability and reduce environmental impact (Xu et al., 2022).

Data analytics is at the core of IoT's contribution to agricultural productivity. The vast amounts of data collected by IoT devices are processed and analyzed to generate actionable insights. Advanced analytics tools and machine learning algorithms can identify patterns and trends that might not be immediately apparent, helping farmers make data-driven decisions. For instance, predictive analytics can forecast weather patterns and their impact on crops, allowing farmers to plan and prepare accordingly. Similarly, data-driven insights into crop performance can guide decisions on planting strategies, crop rotations, and pest management (Raj et al., 2021).

However, the adoption of IoT in agriculture also presents challenges that need to be addressed. The integration of IoT technologies requires significant investment in infrastructure, including sensors, connectivity, and data management systems. Additionally, data security and privacy are critical concerns, as the vast amounts of data collected can be vulnerable to breaches if not properly secured. Ensuring that farmers have the necessary skills and knowledge to use these technologies effectively is also essential for maximizing their benefits. Therefore, a collaborative effort involving technology providers, policymakers, and agricultural stakeholders is crucial to overcome these challenges and promote the widespread adoption of IoT solutions (Armenta-Medina et al., 2020).

In conclusion, advanced IoT solutions are transforming the agricultural sector by enhancing productivity, efficiency, and sustainability. Through precision agriculture, improved crop and livestock management, and better resource utilization, IoT technologies are helping farmers meet the growing global food demand while minimizing environmental impact. As technology continues to evolve, the potential for IoT to drive further advancements in agriculture is substantial. Embracing these innovations and addressing the associated challenges will be key to realizing the full benefits of IoT in agriculture, ensuring a more productive and sustainable future for the industry (Salam, 2024).

Scope

This chapter covers the scope of advanced IoT solutions in agriculture, focusing on the integration of smart technologies to enhance productivity. It includes an in-depth analysis of IoT sensors for monitoring soil health, crop conditions, and environmental variables. The scope extends to discussing precision irrigation systems, automated climate control, and predictive analytics for optimized farming practices. Additionally, the chapter reviews real-world case studies to illustrate successful applications and examines associated challenges, including data security, technology integration, and cost considerations. By addressing these areas, the chapter provides a comprehensive understanding of IoT's potential in transforming modern agriculture.

Objectives

This chapter explores advanced IoT technologies for real-time monitoring of soil conditions, crop health, and environmental factors. It discusses the use of precision irrigation, automated climate control, and predictive analytics to optimize crop yields and resource use. It analyzes case studies of successful IoT implementations in agriculture, addresses challenges like data security, system integration, and cost, and provides actionable insights for researchers, practitioners, and policymakers.

IOT TECHNOLOGIES FOR AGRICULTURE

Figure 1. Advanced IoT technologies in agriculture

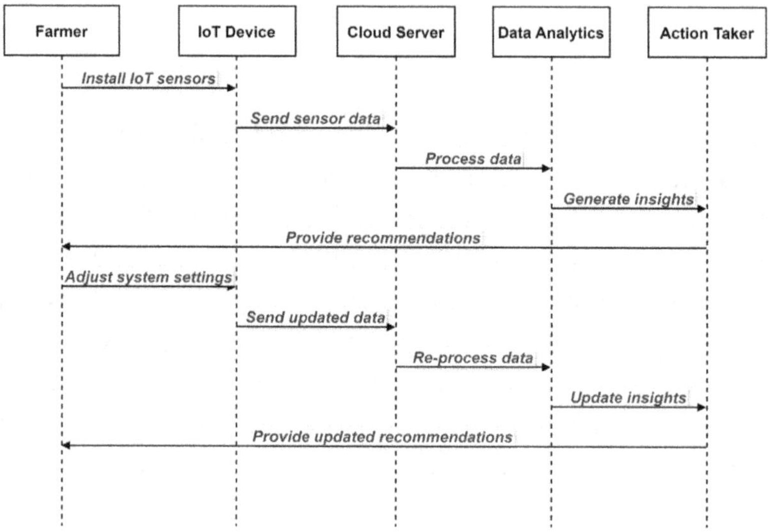

The implementation of advanced IoT technologies in agriculture involves key steps and interactions between various components, as illustrated in Figure 1. The farmer installs IoT sensors in their agricultural field, which collect and send data to a Cloud Server. The data is then processed by Data Analytics, who generates insights and sends them to an Action Taker. The Action Taker then provides recommendations to the farmer, who adjusts the device's settings. The updated data is then sent to the Cloud Server, which re-processes it and sends it to Data Analytics. The insights are then updated and sent to the Action Taker, who provides the farmer with updated recommendations(Jeevanantham et al., 2022a).

Types of IoT Sensors and Devices

In modern agriculture, a variety of IoT sensors and devices play pivotal roles in enhancing productivity and efficiency. These sensors are designed to monitor and collect data on various parameters critical to farming operations.

Soil Moisture Sensors: These devices measure the water content in the soil, providing real-time data on soil moisture levels. They help farmers optimize irrigation schedules, reducing water waste and ensuring that crops receive adequate hydration. Soil moisture sensors come in various types, including capacitive, resistive, and neutron probes, each offering different advantages in terms of accuracy and depth of measurement.

Temperature and Humidity Sensors: These sensors monitor ambient conditions in greenhouses and field environments. Temperature sensors track the heat levels that affect crop growth, while humidity sensors gauge moisture in the air. Maintaining optimal temperature and humidity conditions is crucial for preventing diseases and promoting healthy plant growth.

Nutrient Sensors: These sensors analyze soil and plant tissue for essential nutrients such as nitrogen, phosphorus, and potassium. By assessing nutrient levels, farmers can apply fertilizers more precisely, improving crop yield and reducing excess application, which can lead to environmental pollution.

Weather Stations: Comprehensive weather stations integrate multiple sensors to measure parameters such as temperature, humidity, wind speed, and rainfall. These devices provide valuable data for weather forecasting and climate monitoring, enabling farmers to make informed decisions about planting, harvesting, and managing crops.

Camera Systems: Advanced camera systems, including multispectral and hyperspectral cameras, capture detailed images of crops and soil. These cameras can detect early signs of pests, diseases, and nutrient deficiencies, allowing for timely interventions. Some systems use machine learning algorithms to analyze images and provide actionable insights.

Real-Time Monitoring Systems

Real-time monitoring is a cornerstone of IoT applications in agriculture, allowing for continuous observation and management of farming conditions. These systems integrate various sensors and devices to provide a comprehensive view of agricultural environments (Koshariya, Kalaiyarasi, et al., 2023).

IoT Platforms: Centralized IoT platforms collect data from multiple sensors and devices deployed across the farm. These platforms offer real-time data visualization, enabling farmers to monitor various parameters from a single interface. Dashboards and alerts help users track critical metrics and respond promptly to any issues.
Data Communication Protocols: Real-time data transmission relies on robust communication protocols such as LoRaWAN (Long Range Wide Area Network), Zigbee, and cellular networks (e.g., 4G, 5G). These protocols ensure that data from sensors is transmitted efficiently over long distances, even in remote areas with limited connectivity.
Automated Control Systems: Real-time monitoring systems often include automated control mechanisms. For example, irrigation systems can be automatically adjusted based on soil moisture data, and climate control systems in greenhouses can regulate temperature and humidity based on sensor inputs. Automation reduces the need for manual intervention and enhances operational efficiency.
Integration with Decision Support Systems: Advanced real-time monitoring systems are integrated with decision support systems (DSS) that analyze data and provide actionable recommendations. These systems use algorithms to predict potential issues, such as pest outbreaks or water shortages, and suggest corrective actions.

Data Collection and Integration

Data collection and integration are fundamental to leveraging IoT technologies effectively in agriculture. The process involves gathering, storing, and analyzing data from various sources to inform decision-making and optimize farming practices (Pachiappan et al., 2024).

Data Collection Methods: IoT sensors collect data continuously, providing a wealth of information on soil conditions, weather, and crop health. Data is gathered through wireless communication, often transmitted to central data repositories via cloud-based platforms. This continuous data stream enables real-time monitoring and analysis.

Data Integration: Integrating data from different sensors and sources is crucial for creating a comprehensive view of agricultural conditions. Integration platforms combine data from soil moisture sensors, weather stations, and cameras to provide a unified dataset. This holistic view allows farmers to make more informed decisions based on multiple factors.

Data Analytics: Advanced analytics tools process and analyze the collected data to extract meaningful insights. Machine learning algorithms and statistical models are used to identify patterns, predict future conditions, and recommend actions. For example, data analytics can predict optimal irrigation schedules based on historical and real-time moisture data.

Visualization and Reporting: Data visualization tools present collected data in an accessible format, such as graphs, charts, and dashboards. These visual representations help farmers quickly interpret complex data and identify trends. Reporting tools generate summaries and insights that inform strategic planning and operational adjustments.

Challenges and Solutions: While data collection and integration offer significant benefits, challenges such as data accuracy, sensor calibration, and data security must be addressed. Ensuring the reliability of sensor readings and protecting data from unauthorized access are critical for maintaining the integrity of IoT systems in agriculture.

In summary, IoT technologies in agriculture encompass a range of sensors and devices designed to monitor and enhance various aspects of farming. Real-time monitoring systems provide continuous oversight and automation, while data collection and integration enable comprehensive analysis and informed decision-making. Together, these technologies drive advancements in agricultural productivity, resource efficiency, and overall farm management.

PRECISION IRRIGATION

Principles and Benefits

Principles of Precision Irrigation

Precision irrigation is a sophisticated approach to water management that aims to optimize water usage and improve crop yields by delivering the right amount of water to each plant at the right time (Boopathi, 2024). The core principles of precision irrigation involve:

Targeted Application: Precision irrigation focuses on applying water exactly where it is needed, rather than uniformly across an entire field. This is achieved through detailed analysis of soil moisture, plant water needs, and environmental conditions.
Data-Driven Decisions: The approach relies heavily on data collected from various sources, including soil moisture sensors, weather stations, and satellite imagery. This data informs irrigation decisions and helps adjust water application in real-time based on current conditions.
Efficiency and Minimization of Waste: By applying water more precisely, precision irrigation minimizes water wastage due to evaporation, runoff, and deep percolation. This leads to more efficient water use and reduces the environmental impact of irrigation practices.
Adaptability: Precision irrigation systems can adapt to varying conditions within a field. For example, different soil types, topographies, and crop requirements are taken into account, ensuring that each plant receives the optimal amount of water.

Benefits of Precision Irrigation

Enhanced Crop Yields: By delivering water more effectively, precision irrigation helps crops reach their full potential. Optimized water availability supports better plant growth and can lead to higher yields.

Water Conservation: One of the most significant benefits of precision irrigation is water conservation. By reducing over-irrigation and minimizing wastage, it ensures that water resources are used more sustainably.

Cost Savings: Although the initial investment in precision irrigation systems can be high, the long-term savings are considerable. Reduced water usage and improved crop yields lead to lower operational costs and increased profitability.

Improved Soil Health: Precision irrigation prevents soil erosion and degradation caused by excessive water application. It also helps maintain optimal soil moisture levels, which supports healthier root systems and overall soil quality.

Reduced Environmental Impact: By minimizing water runoff and leaching, precision irrigation helps reduce the risk of pollution and nutrient runoff into water bodies, contributing to better environmental stewardship.

Technologies and Techniques

Technologies in Precision Irrigation

Soil Moisture Sensors: Soil moisture sensors measure the water content in the soil at various depths. These sensors provide real-time data on soil moisture levels, allowing for precise control of irrigation schedules. Types of soil moisture sensors include capacitive sensors, which measure soil dielectric constant, and resistive sensors, which measure electrical resistance (Gnanaprakasam et al., 2023; Jeevanantham et al., 2022b).

Weather Stations: Weather stations collect data on atmospheric conditions such as temperature, humidity, wind speed, and rainfall. This information is crucial for adjusting irrigation practices based on weather forecasts and real-time conditions. Integrated with other data sources, weather stations help in predicting water needs more accurately.

Remote Sensing Technologies: Remote sensing technologies, including satellite imagery and aerial drones, provide valuable insights into crop health and soil conditions. Multispectral and hyperspectral images can identify variations in plant stress and soil moisture levels across large areas, guiding targeted irrigation practices.
Automated Irrigation Systems: Automated irrigation systems use data from sensors and weather stations to control irrigation equipment. These systems include drip irrigation, which delivers water directly to plant roots through a network of tubes and emitters, and sprinkler systems, which can be programmed to adjust water application based on real-time data.
Flow Meters and Valves: Flow meters measure the amount of water being delivered to different areas of a field. Combined with electronically controlled valves, these devices enable precise regulation of water flow, ensuring that each section of the field receives the appropriate amount of water.

Techniques in Precision Irrigation

Drip Irrigation: Drip irrigation involves delivering water directly to the root zone of plants through a network of tubes and emitters. This technique minimizes water wastage and ensures that each plant receives a consistent supply of water. Drip irrigation is particularly effective for row crops and high-value plants (Gnanaprakasam et al., 2023; Jeya et al., 2023; Sankar et al., 2023).
Sprinkler Irrigation: Modern sprinkler systems can be equipped with sensors and controllers to adjust water application based on real-time conditions. Techniques such as center pivot irrigation and variable rate irrigation allow for more precise water distribution over large areas.
Subsurface Drip Irrigation (SDI): SDI involves placing drip lines below the soil surface, delivering water directly to the root zone while reducing evaporation and surface runoff. This technique is suitable for deep-rooted crops and areas with high evaporation rates.

Micro-Irrigation: Micro-irrigation systems, including micro-sprinklers and misting systems, provide small quantities of water to specific areas. These systems are ideal for delicate plants and high-value crops, offering precise control over water application.

Variable Rate Irrigation (VRI): VRI technology adjusts the amount of water applied to different areas of a field based on data from sensors and remote sensing technologies. This technique allows for targeted irrigation, optimizing water use and improving crop performance.

In summary, precision irrigation leverages advanced technologies and techniques to deliver water more efficiently and effectively. By adhering to the principles of targeted application, data-driven decisions, and adaptability, precision irrigation systems offer numerous benefits, including enhanced crop yields, water conservation, cost savings, and reduced environmental impact. The integration of soil moisture sensors, weather stations, remote sensing technologies, automated systems, and advanced irrigation techniques ensures that water resources are utilized to their fullest potential, supporting sustainable and productive agricultural practices.

AUTOMATED CLIMATE CONTROL

Smart Climate Management Systems

Automated climate control systems in agriculture are designed to optimize the environmental conditions within greenhouses, indoor farms, and controlled-environment agriculture systems. These systems leverage advanced technologies to regulate temperature, humidity, light, and air circulation, ensuring optimal growing conditions for crops(Hema et al., 2023; Koshariya, Kalaiyarasi, et al., 2023; S., 2024).

Figure 2. Smart climate management system for agriculture

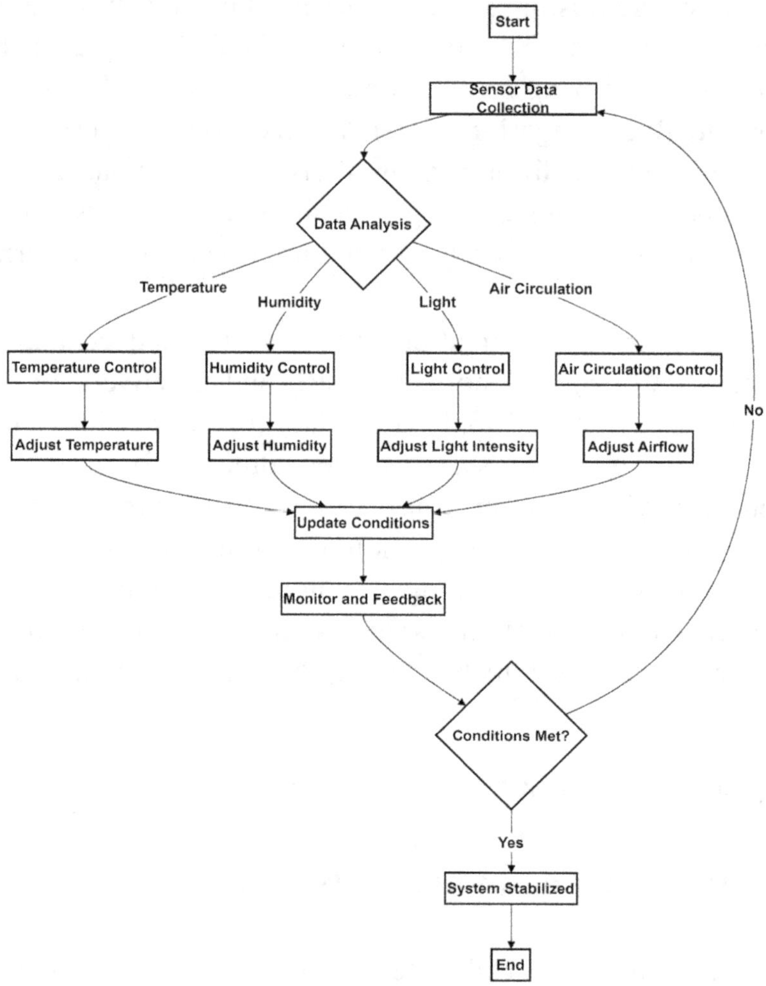

Figure 2 illustrates a Smart Climate Management System for agriculture, outlining key components and interactions for automating climate control. The process of a Smart Climate Management System begins with sensor data collection, which is then analyzed to determine the need for adjustments. The system then controls temperature, humidity, light intensity, and air circulation. Adjustments are made to these settings, ensuring optimal environmental conditions. The system then updates the conditions based on these adjustments. It continuously monitors the environment and provides feedback, checking if the conditions are now optimal. If these conditions are met, the system stabilizes, and the process concludes. This diagram provides a high-level overview of how a Smart Climate Management System operates.

Components of Smart Climate Management Systems

Climate Sensors: These sensors continuously monitor key environmental parameters such as temperature, humidity, CO2 levels, and light intensity. They provide real-time data that is crucial for maintaining ideal growing conditions. Sensors are often integrated with other system components to ensure precise control (Koshariya, Kalaiyarasi, et al., 2023; Koshariya, Khatoon, et al., 2023).

HVAC Systems: Heating, Ventilation, and Air Conditioning (HVAC) systems are central to automated climate control. They adjust the temperature and humidity levels based on data from climate sensors. Advanced HVAC systems can be programmed to respond to specific conditions, such as increasing ventilation when humidity levels are too high or heating when temperatures drop below a certain threshold.

Lighting Systems: Artificial lighting is essential in controlled environments, especially in greenhouses where natural light may be insufficient. Smart lighting systems use LED technology and can be adjusted based on the time of day, plant growth stages, and real-time light intensity measurements. They provide the necessary light spectrum and intensity to support photosynthesis and plant development.

Irrigation and Fertilization Integration: Climate control systems often integrate with irrigation and fertilization systems. Automated climate control can adjust irrigation schedules based on humidity levels and plant needs, ensuring that plants receive the right amount of water and nutrients.

Climate Control Software: Centralized software platforms manage and analyze data from various sensors and control systems. They provide a user interface for monitoring and adjusting environmental conditions. Advanced software includes predictive analytics and machine learning algorithms to anticipate changes and optimize climate settings proactively.

Control Interfaces: These interfaces allow for manual adjustments and monitoring of the climate control system. They can be accessed remotely via mobile apps or web interfaces, enabling farmers and operators to manage their climate control systems from anywhere.

Benefits of Smart Climate Management Systems

Optimal Growing Conditions: Automated climate control systems maintain ideal conditions for plant growth, regardless of external weather conditions. This results in healthier plants, more consistent growth, and potentially higher yields.
Resource Efficiency: By precisely managing temperature, humidity, and lighting, these systems reduce energy consumption and water usage. This leads to lower operational costs and more sustainable farming practices.
Improved Crop Quality: Consistent environmental conditions contribute to uniform crop quality. Automated systems reduce the risk of plant stress and variability, leading to higher-quality produce.
Increased Productivity: By optimizing growing conditions and reducing manual labor, automated climate control systems can increase overall productivity. Farmers can achieve more efficient use of space and resources, leading to better returns on investment.
Data-Driven Insights: The integration of climate sensors and software provides valuable data and insights into plant growth and environmental conditions. This data can be used to refine growing practices and make informed decisions.

Impact on Crop Health and Yield

Enhanced Crop Health

Automated climate control systems reduce plant stress by maintaining consistent temperature, humidity, and light levels, promoting plant health and reducing susceptibility to diseases and pests. They also prevent diseases by regulating humidity and temperature, preventing mold and mildew growth. Smart lighting systems regulate light intensity and duration, promoting efficient photosynthesis and overall plant health (Agrawal, Magulur, et al., 2023; Agrawal, Shashibhushan, et al., 2023; Gnanaprakasam et al., 2023; Koshariya, Kalaiyarasi, et al., 2023). Climate control systems ensure plants receive the right amount of water and nutrients, preventing overwatering or imbalances, supporting healthy root development and nutrient uptake.

Figure 3. Influence of automated climate control systems on crop health and yield

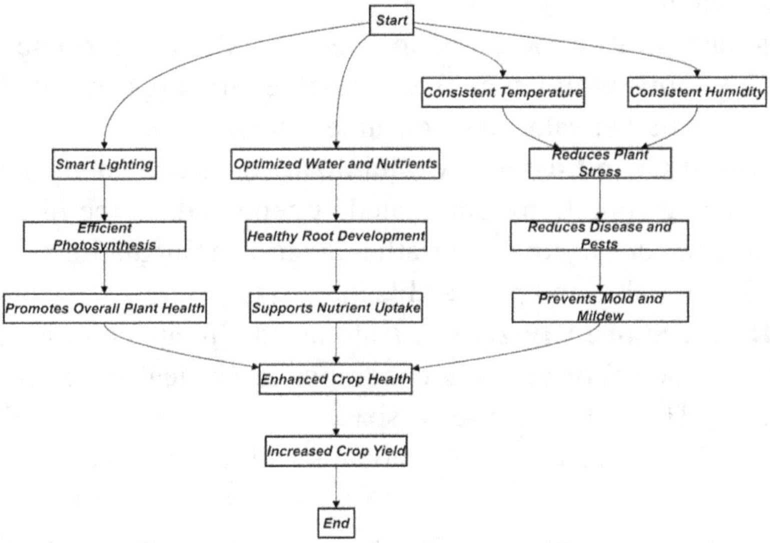

The figure 3 demonstrates the influence of automated climate control systems on crop health and yield. Automated climate control systems are crucial for maintaining crop health and increasing yield by regulating temperature, humidity, lighting, water and nutrient supply, reducing plant stress, improving photosynthesis, supporting healthy root development, reducing disease and pests, promoting overall plant health, supporting nutrient uptake, preventing mold and mildew, and enhancing crop health. These systems help maintain optimal conditions, reduce risks, and ultimately lead to better crop yields. The process concludes with a flowchart that illustrates how these systems contribute to improved crop health and increased yield (Boopathi, 2024; Glady et al., 2024; Pachiappan et al., 2024; Venkateswaran et al., 2024).

Increased Yield

Consistent Growth Rates: By providing uniform environmental conditions, automated climate control systems ensure that plants grow at a consistent rate. This leads to more uniform harvests and maximizes crop yield potential.

Extended Growing Seasons: In controlled environments, crops can be grown year-round, independent of external weather conditions. This enables multiple growing cycles per year and increases overall production.

Higher Quality Produce: Consistent conditions contribute to higher quality crops with better flavor, texture, and appearance. This can enhance market value and consumer satisfaction.

Reduced Crop Losses: By minimizing environmental stresses and optimizing conditions, automated systems reduce the likelihood of crop losses due to adverse weather or climate fluctuations. This leads to more reliable and predictable harvests.

Efficient Space Utilization: Automated climate control allows for precise control of growing conditions in vertical farms and greenhouses. This efficient use of space can result in higher yields per square foot compared to traditional farming methods.

In summary, automated climate control systems are transformative tools in modern agriculture, offering enhanced management of environmental conditions. By leveraging advanced technologies and integrating various components, these systems optimize growing conditions, leading to improved crop health, higher yields, and greater resource efficiency. The benefits of smart climate management extend beyond individual farms, contributing to more sustainable and productive agricultural practices.

PREDICTIVE ANALYTICS IN AGRICULTURE

Figure 4. Predictive analytics in agriculture

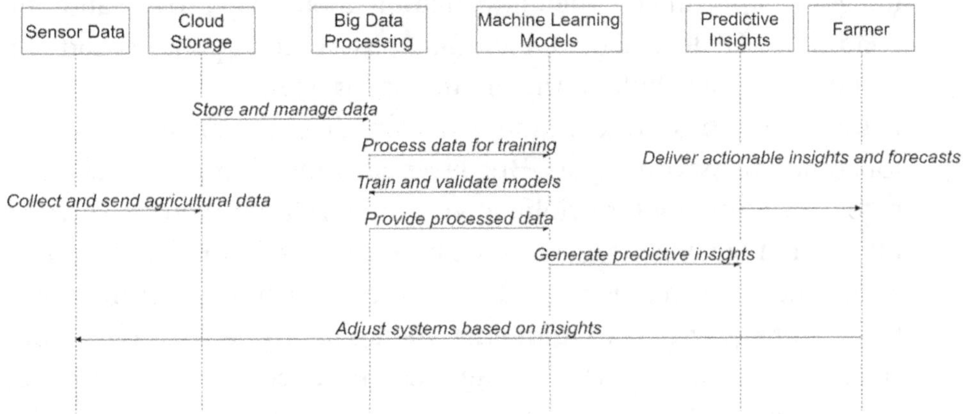

The process of leveraging Big Data and Machine Learning for predictive analytics in agriculture involves collecting and sending sensor data to Cloud Storage, managing and storing large volumes of data, handling big data processing, training and validating machine learning models, generating predictive insights, providing farmers with actionable insights to make informed decisions, and adjusting agricultural systems based on these insights. Figure 4 illustrates the flow of data and interactions involved in this process.

Role of Big Data and Machine Learning

Big Data in Agriculture

Big data refers to the vast volumes of data generated from various sources within agricultural operations. In agriculture, big data encompasses information from sensors, weather stations, satellite imagery, and even social media. This data provides valuable insights into soil health, crop performance, and environmental conditions.

Data Collection: Modern farming techniques utilize a wide range of data sources. Sensors collect real-time data on soil moisture, temperature, and crop health. Drones and satellites provide high-resolution imagery, capturing detailed information about crop growth and field conditions. Historical data, including past weather patterns and crop yields, also contribute to the big data ecosystem.

Data Integration: Integrating data from diverse sources is essential for comprehensive analysis. Big data platforms aggregate and harmonize data, allowing for a holistic view of agricultural processes. This integration enables farmers to see how various factors interact, such as the impact of weather conditions on soil moisture and crop growth.

Data Storage and Management: Managing big data requires robust storage solutions and data management practices. Cloud-based storage systems and big data platforms handle the large volumes of data generated in agriculture. These systems ensure that data is securely stored and easily accessible for analysis.

Machine Learning in Agriculture

Machine learning (ML) algorithms analyze large datasets to identify patterns and make predictions. In agriculture, ML enhances predictive analytics by providing insights that drive decision-making and optimize farming practices.

Pattern Recognition: ML algorithms detect patterns and correlations within data that may not be immediately obvious. For example, ML can identify patterns in weather data and soil conditions that correlate with pest outbreaks or crop diseases.

Predictive Modeling: ML models predict future outcomes based on historical data. For instance, predictive models can forecast crop yields by analyzing historical weather patterns, soil health, and planting practices. These predictions help farmers make informed decisions about resource allocation and management.

Automation and Optimization: ML algorithms automate repetitive tasks and optimize processes. For example, ML-driven irrigation systems can automatically adjust water schedules based on real-time data, optimizing water use and reducing waste.

Tools and Techniques for Data Analysis

Data Analytics Platforms: Specialized platforms and software tools are used to analyze agricultural data. Platforms like IBM Watson, Microsoft Azure, and Google Cloud offer machine learning and analytics capabilities tailored to agricultural applications. These tools provide data visualization, reporting, and predictive modeling features.

Geographic Information Systems (GIS): GIS technology integrates spatial data with other types of information to analyze geographic patterns. In agriculture, GIS helps visualize and analyze spatial data related to soil types, crop health, and field variability. This spatial analysis supports precision agriculture practices and decision-making.

Data Visualization Tools: Visualization tools such as Tableau, Power BI, and Qlik Sense help present complex data in a user-friendly format. These tools create interactive dashboards, charts, and maps that make it easier for farmers and analysts to interpret data and identify trends.

Statistical Analysis Software: Statistical tools like R and Python libraries (e.g., Pandas, SciPy) are used for in-depth data analysis and model building. These tools support various statistical techniques, including regression analysis, clustering, and hypothesis testing, which are essential for developing predictive models.

Remote Sensing Technology: Remote sensing technologies, including satellite imagery and drones, capture data on crop health, soil conditions, and environmental factors. Remote sensing data is analyzed to monitor crop growth, assess field conditions, and detect early signs of stress or disease.

IoT and Sensor Networks: IoT devices and sensor networks provide real-time data on soil moisture, temperature, and other environmental conditions. Analyzing data from these sensors enables precise monitoring and management of agricultural processes, such as irrigation and fertilization.

Predictive Models and Their Applications

Yield Prediction Models: Predictive models forecast crop yields based on historical data and current conditions. These models analyze factors such as weather patterns, soil health, and crop management practices. Accurate yield predictions help farmers plan harvests, manage resources, and make marketing decisions.
Disease and Pest Forecasting Models: Machine learning models predict the likelihood of pest outbreaks and disease occurrences by analyzing historical data, weather conditions, and crop health indicators. Early predictions enable farmers to implement preventive measures and reduce crop losses.
Irrigation Scheduling Models: Predictive models optimize irrigation schedules based on real-time data from soil moisture sensors, weather forecasts, and crop water requirements. These models help farmers apply the right amount of water at the right time, improving water efficiency and crop health.
Fertilization Models: Predictive analytics models assess soil nutrient levels and recommend optimal fertilization practices. By analyzing soil data and crop requirements, these models ensure that fertilizers are applied efficiently, enhancing crop growth and minimizing environmental impact.
Risk Management Models: Predictive models assess various risks, such as weather extremes, market fluctuations, and supply chain disruptions. By analyzing historical data and current trends, these models help farmers develop strategies to mitigate risks and adapt to changing conditions.
Precision Agriculture Models: These models use data from sensors, remote sensing technologies, and other sources to optimize farming practices at a granular level. Precision agriculture models enable site-specific management of crops, improving resource use efficiency and overall productivity.

Predictive analytics is a vital tool in modern agriculture, utilizing big data and machine learning to optimize farming practices. Tools like data analytics platforms, GIS, and remote sensing technology aid in developing predictive models, providing valuable insights into crop yields, disease forecasting, irrigation scheduling, and productivity enhancement.

CASE STUDIES

Successful IoT applications in agriculture demonstrate the transformative potential of technology in enhancing productivity, efficiency, and sustainability. From precision agriculture and smart irrigation to livestock monitoring and greenhouse management, IoT technologies are enabling farmers to make data-driven decisions and optimize their practices. The impact of these technologies includes improved resource management, increased yields, reduced costs, and enhanced food safety. As IoT continues to evolve, its applications in agriculture are expected to expand, offering new opportunities for innovation and growth in the industry (Agrawal, Magulur, et al., 2023; Agrawal, Shashibhushan, et al., 2023; Sankar et al., 2023).

Precision Agriculture

Precision agriculture involves using IoT technologies to optimize farming practices by collecting and analyzing data on soil conditions, crop health, and environmental factors. This approach enables farmers to make informed decisions and apply resources more efficiently. John Deere has integrated IoT technology into its farming equipment to enhance precision agriculture. Their systems use sensors and GPS technology to monitor soil conditions and crop health in real-time. For example, their tractors and combine harvesters are equipped with IoT sensors that provide data on soil moisture, nutrient levels, and crop yield. This data helps farmers adjust their practices, such as varying seed planting rates and fertilizer application based on specific field conditions, leading to optimized resource use and increased crop yields (Agrawal, Magulur, et al., 2023; Boopathi & Khang, 2023).

Impact: John Deere's precision agriculture solutions have led to more efficient use of inputs, reduced waste, and improved crop yields. Farmers using these technologies have reported up to a 20% increase in crop production and significant savings in water and fertilizer costs.

Smart Irrigation Systems

Smart irrigation systems use IoT sensors to monitor soil moisture levels and weather conditions, allowing for automated and optimized irrigation schedules. These systems ensure that crops receive the right amount of water, reducing waste and enhancing crop health.

The California Almond Board implemented smart irrigation systems to address water scarcity issues in almond orchards. The project involved installing soil moisture sensors and weather stations throughout the orchards. The data collected by these sensors is transmitted to a central platform that analyzes soil moisture levels and weather forecasts to determine the optimal irrigation schedule.

Impact: The smart irrigation systems have led to a 15-25% reduction in water usage while maintaining or improving almond yields. By optimizing irrigation practices, the project has also contributed to more sustainable water management in the region.

Livestock Monitoring

Application Overview: IoT technologies are used in livestock monitoring to track the health and well-being of animals. Sensors can monitor vital signs, activity levels, and environmental conditions to provide insights into livestock health and behavior.

Allflex offers a range of IoT solutions for livestock management, including wearable sensors that track animal health and behavior. These sensors monitor parameters such as body temperature, activity levels, and feeding patterns. Data is collected and analyzed to detect early signs of illness, optimize feeding schedules, and manage breeding programs.

Impact: Farmers using Allflex's livestock monitoring solutions have reported improvements in animal health and productivity. Early detection of health issues has led to reduced veterinary costs and increased overall herd performance. The technology also helps in making informed decisions about feeding and breeding, contributing to better management practices.

Greenhouse Management

In greenhouse settings, IoT technologies are used to manage environmental conditions, such as temperature, humidity, and light levels, to create optimal growing conditions for plants. Smart greenhouses use sensors and automated systems to regulate these factors.

Netafim has developed a smart greenhouse solution that integrates IoT sensors, climate control systems, and data analytics. Sensors monitor various environmental parameters, including temperature, humidity, CO2 levels, and light intensity. The data is used to control climate control systems, such as ventilation, heating, and irrigation, ensuring that plants grow under ideal conditions.

Impact: Netafim's smart greenhouse solutions have led to increased crop yields and reduced energy and water consumption. By maintaining optimal growing conditions, the technology has also enhanced the quality of produce and reduced the reliance on manual labor for climate management.

Supply Chain Tracking

IoT technologies are used to track agricultural products throughout the supply chain, from farm to table. Sensors and RFID tags provide real-time data on location, temperature, and handling conditions, ensuring product quality and traceability. IBM's Food Trust blockchain platform, in collaboration with Walmart, uses IoT and blockchain technology to track the journey of food products through the supply chain. RFID tags and sensors are used to monitor the conditions under which food is stored and transported. The data is recorded on a blockchain ledger, providing a transparent and immutable record of the product's journey.

Impact: The use of IoT for supply chain tracking has improved transparency, reduced food waste, and enhanced food safety. Walmart has reported faster traceability of food products, enabling quicker responses to food safety issues and ensuring that consumers receive high-quality products.

Pest and Disease Management

IoT technologies are employed to monitor pest populations and disease outbreaks in crops. Sensors and imaging technologies detect early signs of pest infestations and plant diseases, allowing for timely interventions. Microsoft's FarmBeats project leverages IoT sensors, drones, and machine learning to monitor crop health and detect pest infestations. The system uses sensors to collect data on soil conditions, crop growth, and pest activity. Drones capture aerial images, which are analyzed to identify early signs of pest problems or diseases.

Impact: FarmBeats has helped farmers reduce pesticide use and increase crop yields by enabling more targeted and timely interventions. Early detection of pests and diseases has led to improved crop health and reduced losses, contributing to more sustainable farming practices.

CHALLENGES AND SOLUTIONS

Data Security Concerns

Challenges

Agricultural data, including sensitive information about crop yields, farm operations, and financial details, is vulnerable to cyber-attacks, compromising farm security. Ensuring data privacy is crucial, as it often includes personal information about farmers. Unauthorized access to data increases as predictive analytics systems become interconnected, leading to potential data tampering or theft. Ensuring data protection regulations and preventing data breaches is essential for maintaining privacy and compliance in agricultural systems (Domakonda et al., 2022; Samikannu et al., 2022; Vennila et al., 2022).

Solutions

Encryption is crucial for protecting sensitive information from unauthorized access, ensuring it remains unreadable even if intercepted or accessed without permission. Access controls, such as multi-factor authentication and role-based access control, ensure only authorized personnel can access and manage data. Regular security audits and vulnerability assessments help identify system weaknesses, while updates and patches prevent exploitation of known vulnerabilities. Data anonymization, which involves removing or obfuscating personally identifiable information, protects personal information and reduces privacy risks by removing or obfuscating PII.

Integration with Existing Systems

Challenges

Integrating new predictive analytics tools with existing agricultural management systems can be challenging due to compatibility issues, data silos, and complex integration processes. Incompatibilities can hinder seamless data exchange and system functionality. Data silos can limit predictive analytics effectiveness by restricting access to relevant data. The integration process can be complex, requiring customization and adaptation of both new and existing systems, leading to delays, increased costs, and potential operational disruptions.

Solutions

Standardized data formats like XML or JSON can facilitate seamless integration between different systems, reducing challenges. APIs, which allow seamless communication and data sharing, streamline processes and ensure predictive analytics tools can effectively interact with existing systems. Middleware solutions act as intermediaries, providing a unified interface for data exchange and processing. Engaging with experts and training staff on new systems and integration processes can provide valuable insights and support, ensuring a smooth transition and minimizing disruptions. Overall, adopting these strategies can streamline the integration process and improve overall system performance.

Cost Implications and Budgeting

Challenges

Predictive analytics solutions often require a high initial investment, which can be a barrier for smaller farms with limited budgets. Ongoing maintenance costs, such as software updates, hardware repairs, and system support, can further burden the financial burden, impacting the overall budget and profitability of agricultural operations. Demonstrating the ROI of predictive analytics can be challenging, as the benefits may take time to materialize and quantifying the financial impact can be challenging.

Solutions

A cost-benefit analysis is crucial for evaluating the financial implications of implementing predictive analytics in agriculture. It should consider both initial investment and potential long-term benefits like increased yields and resource savings. A phased implementation approach can help spread costs over time and evaluate the system's effectiveness before full-scale deployment. Grants, subsidies, and financial incentives can offset costs, and government programs and industry initiatives can provide financial support. Scalable solutions can help manage costs and adapt to changing needs, avoiding unnecessary expenses. Partnering with technology vendors and service providers can offer financial benefits, discounts, payment plans, and technical support, as well as access to valuable resources and expertise.

In summary, addressing challenges related to data security, system integration, and cost implications is crucial for successfully implementing predictive analytics in agriculture. By adopting robust security measures, ensuring compatibility with existing systems, and carefully managing costs, farms can leverage predictive analytics to enhance productivity, optimize resource use, and achieve better outcomes. Effective solutions and strategic planning can mitigate challenges and maximize the benefits of predictive analytics, driving innovation and growth in the agricultural sector.

FUTURE TRENDS AND DIRECTIONS

Figure 5. Future trends and directions in IOT for agricultural activities

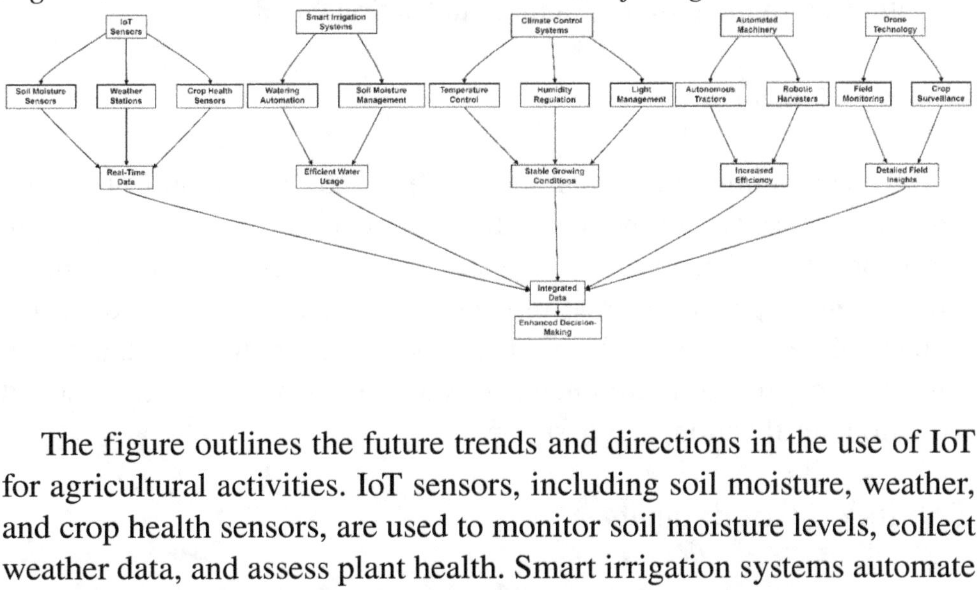

The figure outlines the future trends and directions in the use of IoT for agricultural activities. IoT sensors, including soil moisture, weather, and crop health sensors, are used to monitor soil moisture levels, collect weather data, and assess plant health. Smart irrigation systems automate watering and manage soil moisture levels. Climate control systems regulate environmental conditions, such as temperature, humidity, and light levels. Automated machinery, including autonomous tractors and robotic harvesters, perform field work. Drone technology provides field monitoring and crop surveillance. Data analytics integrates and analyzes data from various sources, providing real-time information for efficient water usage. This technology enhances machinery performance, provides detailed field insights, and enhances decision-making for better agricultural decisions. Overall, IoT technology is transforming agriculture (Agrawal, Magulur, et al., 2023; Boopathi, 2024; Koshariya, Kalaiyarasi, et al., 2023; Pachiappan et al., 2024; Venkateswaran et al., 2024).

Emerging IoT Technologies

Advanced Sensor Technology

Emerging IoT technologies are significantly enhancing the capabilities of agricultural sensors. Next-generation sensors are increasingly sophisticated, offering improved accuracy, durability, and integration with other technologies. These advanced sensors can monitor a wider range of parameters, including soil health, plant stress indicators, and microclimate conditions, providing more granular and actionable data.

Edge Computing

Edge computing is becoming a crucial trend in IoT applications for agriculture. Instead of sending all data to a centralized cloud, edge computing processes data locally on devices or nearby servers. This approach reduces latency, enhances real-time decision-making, and decreases the amount of data transmitted over networks. Edge computing allows for quicker responses to environmental changes, improving the efficiency of irrigation, fertilization, and pest control.

Enhanced Connectivity

Emerging connectivity technologies, such as 5G and Low Power Wide Area Networks (LPWAN), are set to revolutionize IoT in agriculture. 5G offers high-speed, low-latency communication, which is essential for real-time data processing and remote control of agricultural systems. LPWAN technologies like LoRaWAN (Long Range Wide Area Network) provide long-range, low-power communication for sensor networks, making them ideal for large agricultural operations.

Integration of AI with IoT

Artificial Intelligence (AI) and IoT are converging to create more intelligent and autonomous agricultural systems. AI algorithms analyze data from IoT sensors to predict crop health, optimize resource use, and automate tasks such as irrigation and pest control. This integration enhances the precision and efficiency of agricultural practices, leading to more sustainable and productive farming.

Innovations on the Horizon

Autonomous Farming Equipment

Autonomous farming equipment, such as drones, tractors, and harvesters, is on the rise. These machines use AI and IoT technologies to operate independently, performing tasks such as planting, weeding, and harvesting with minimal human intervention. Autonomous equipment increases operational efficiency, reduces labor costs, and allows for precise execution of agricultural tasks.

Precision Agriculture Platforms

Innovative precision agriculture platforms are emerging, offering integrated solutions for monitoring, analyzing, and managing agricultural operations. These platforms combine data from multiple sources, including IoT sensors, satellite imagery, and weather forecasts, to provide comprehensive insights and recommendations. They enable farmers to make data-driven decisions and optimize farming practices across various aspects, including irrigation, fertilization, and pest management.

Advanced Data Analytics Tools

The development of advanced data analytics tools is transforming how agricultural data is analyzed and utilized. Machine learning models and predictive analytics algorithms are becoming more sophisticated, enabling deeper insights and more accurate predictions. These tools help farmers anticipate and respond to challenges such as weather events, pest outbreaks, and soil degradation.

Smart Greenhouses

Smart greenhouses are an area of rapid innovation, integrating IoT sensors, climate control systems, and AI-driven analytics to optimize growing conditions. These greenhouses can automatically adjust temperature, humidity, and light levels based on real-time data, creating an ideal environment for plant growth. Innovations in smart greenhouse technology aim to improve energy efficiency, reduce resource use, and enhance crop yields.

Blockchain for Agricultural Supply Chains

Blockchain technology is being explored for improving transparency and traceability in agricultural supply chains. By providing a decentralized and immutable ledger of transactions, blockchain can enhance food safety, reduce fraud, and ensure the authenticity of agricultural products. This innovation offers potential benefits in terms of quality control, traceability, and accountability.

Potential Impact on Agricultural Practices

Improved Resource Management

Emerging IoT technologies and innovations will lead to more efficient management of resources, such as water, nutrients, and energy. Real-time monitoring and predictive analytics enable precise application of resources, reducing waste and minimizing environmental impact. For example, smart irrigation systems can optimize water use based on soil moisture data and weather forecasts, conserving water and improving crop health.

Enhanced Crop Yields

The integration of advanced technologies in agriculture is expected to boost crop yields by optimizing growing conditions and minimizing losses. Precision agriculture and autonomous equipment enable more accurate planting, fertilization, and pest control, leading to healthier plants and higher yields. Innovations in data analytics also contribute to better decision-making and improved crop management.

Increased Sustainability

Innovations in IoT and data analytics support sustainable farming practices by reducing resource consumption, minimizing environmental impact, and promoting efficient use of inputs. Technologies such as smart greenhouses and autonomous equipment help farmers adopt more sustainable practices, including reduced pesticide use, optimized water management, and lower carbon emissions.

Greater Operational Efficiency

The automation and integration of advanced technologies streamline agricultural operations, reducing labor requirements and increasing productivity. Autonomous equipment, precision agriculture platforms, and real-time data analysis enable more efficient management of farm activities, leading to cost savings and improved operational efficiency.

Enhanced Resilience to Climate Change

Predictive analytics and advanced IoT technologies enhance resilience to climate change by providing accurate forecasts and early warnings of extreme weather events. Farmers can use this information to adapt their practices and mitigate the impact of climate-related challenges. For instance, predictive models can help farmers prepare for droughts, floods, or temperature fluctuations, reducing risks and improving crop resilience.

In summary, the future of predictive analytics in agriculture is shaped by emerging IoT technologies, innovative solutions, and their potential impacts on farming practices. Advanced sensor technology, edge computing, enhanced connectivity, and AI integration are driving significant advancements. Innovations such as autonomous farming equipment, precision agriculture platforms, and smart greenhouses are set to transform agricultural operations, leading to improved resource management, increased yields, sustainability, and operational efficiency. As these technologies continue to evolve, they promise to revolutionize the agricultural industry, making it more resilient, productive, and sustainable.

CONCLUSION

The integration of IoT solutions in agriculture is revolutionizing productivity, sustainability, and efficiency. As global food demand increases and environmental challenges become more pressing, IoT technologies play a crucial role in optimizing agricultural operations. Precision agriculture, driven by IoT sensors and data analytics, improves crop yields and reduces waste. Smart irrigation systems conserve water and maintain crop health, while livestock monitoring solutions offer real-time insights. IoT-enabled climate control optimizes growing conditions and resource use, while supply chain tracking ensures food safety and quality. The adoption of IoT technologies fundamentally alters agricultural practices by introducing greater efficiency and precision, reducing reliance on traditional methods and minimizing environmental footprint.

The future of IoT in agriculture is promising, with advancements in edge computing, 5G, and artificial intelligence enhancing its capabilities. These technologies will enable farmers to make informed decisions, optimize resource use, and adapt to changing conditions more efficiently. However, implementing IoT solutions in agriculture faces challenges such as data security, system integration, and cost implications. To fully realize these technologies' potential, robust cybersecurity measures, seamless integration with existing systems, and effective financial management are crucial for successful implementation and long-term sustainability.

Hence, advanced IoT solutions are reshaping the agricultural landscape by providing innovative tools and insights that drive productivity and sustainability. The continued development and adoption of these technologies promise to further transform the industry, making agriculture more efficient, responsive, and environmentally friendly. As we move forward, it is essential for stakeholders—farmers, technology providers, and policymakers—to collaborate in harnessing the full potential of IoT, overcoming challenges, and fostering an agricultural sector that meets the demands of a growing global population while preserving natural resources for future generations.

REFERENCES

Agrawal, A. V., Magulur, L. P., Priya, S. G., Kaur, A., Singh, G., & Boopathi, S. (2023). Smart Precision Agriculture Using IoT and WSN. In *Handbook of Research on Data Science and Cybersecurity Innovations in Industry 4.0 Technologies* (pp. 524–541). IGI Global. DOI: 10.4018/978-1-6684-8145-5.ch026

Agrawal, A. V., Shashibhushan, G., Pradeep, S., Padhi, S., Sugumar, D., & Boopathi, S. (2023). Synergizing Artificial Intelligence, 5G, and Cloud Computing for Efficient Energy Conversion Using Agricultural Waste. In *Sustainable Science and Intelligent Technologies for Societal Development* (pp. 475–497). IGI Global.

Armenta-Medina, D., Ramirez-delReal, T. A., Villanueva-Vásquez, D., & Mejia-Aguirre, C. (2020). Trends on advanced information and communication technologies for improving agricultural productivities: A bibliometric analysis. *Agronomy (Basel), 10*(12), 1989. DOI: 10.3390/agronomy10121989

Boopathi, S. (2024). Sustainable Development Using IoT and AI Techniques for Water Utilization in Agriculture. In *Sustainable Development in AI, Blockchain, and E-Governance Applications* (pp. 204–228). IGI Global. DOI: 10.4018/979-8-3693-1722-8.ch012

Boopathi, S., & Khang, A. (2023). AI-Integrated Technology for a Secure and Ethical Healthcare Ecosystem. In *AI and IoT-Based Technologies for Precision Medicine* (pp. 36–59). IGI Global. DOI: 10.4018/979-8-3693-0876-9.ch003

Dhanaraju, M., Chenniappan, P., Ramalingam, K., Pazhanivelan, S., & Kaliaperumal, R. (2022). Smart farming: Internet of Things (IoT)-based sustainable agriculture. *Agriculture, 12*(10), 1745. DOI: 10.3390/agriculture12101745

Domakonda, V. K., Farooq, S., Chinthamreddy, S., Puviarasi, R., Sudhakar, M., & Boopathi, S. (2022). Sustainable Developments of Hybrid Floating Solar Power Plants: Photovoltaic System. In *Human Agro-Energy Optimization for Business and Industry* (pp. 148–167). IGI Global.

Glady, J. B. P., D'Souza, S. M., Priya, A. P., Amuthachenthiru, K., Vikram, G., & Boopathi, S. (2024). A Study on AI-ML-Driven Optimizing Energy Distribution and Sustainable Agriculture for Environmental Conservation. In *Harnessing High-Performance Computing and AI for Environmental Sustainability* (pp. 1–27). IGI Global., DOI: 10.4018/979-8-3693-1794-5.ch001

Gnanaprakasam, C., Vankara, J., Sastry, A. S., Prajval, V., Gireesh, N., & Boopathi, S. (2023). Long-Range and Low-Power Automated Soil Irrigation System Using Internet of Things: An Experimental Study. In *Contemporary Developments in Agricultural Cyber-Physical Systems* (pp. 87–104). IGI Global.

Hema, N., Krishnamoorthy, N., Chavan, S. M., Kumar, N., Sabarimuthu, M., & Boopathi, S. (2023). A Study on an Internet of Things (IoT)-Enabled Smart Solar Grid System. In *Handbook of Research on Deep Learning Techniques for Cloud-Based Industrial IoT* (pp. 290–308). IGI Global. DOI: 10.4018/978-1-6684-8098-4.ch017

Jeevanantham, Y. A., Saravanan, A., Vanitha, V., Boopathi, S., & Kumar, D. P. (2022a). Implementation of Internet-of Things (IoT) in Soil Irrigation System. *IEEE Explore*, 1–5.

Jeevanantham, Y. A., Saravanan, A., Vanitha, V., Boopathi, S., & Kumar, D. P. (2022b). Implementation of Internet-of Things (IoT) in Soil Irrigation System. *IEEE Explore*, 1–5.

Jeya, R., Venkatakrishnan, G. R., Rengaraj, R., Rajalakshmi, M., Pradeep Mohan Kumar, K., & Boopathi, S. (2023). Water Resource Managements in Soil and Soilless Irrigation Systems Using AI Techniques. In *Advances in Environmental Engineering and Green Technologies* (pp. 245–266). IGI Global. DOI: 10.4018/979-8-3693-0338-2.ch014

Koshariya, A. K., Kalaiyarasi, D., Jovith, A. A., Sivakami, T., Hasan, D. S., & Boopathi, S. (2023). AI-Enabled IoT and WSN-Integrated Smart Agriculture System. In *Artificial Intelligence Tools and Technologies for Smart Farming and Agriculture Practices* (pp. 200–218). IGI Global. DOI: 10.4018/978-1-6684-8516-3.ch011

Koshariya, A. K., Khatoon, S., Marathe, A. M., Suba, G. M., Baral, D., & Boopathi, S. (2023). Agricultural Waste Management Systems Using Artificial Intelligence Techniques. In *AI-Enabled Social Robotics in Human Care Services* (pp. 236–258). IGI Global. DOI: 10.4018/978-1-6684-8171-4.ch009

Mishra, S., & Sharma, S. K. (2023). Advanced contribution of IoT in agricultural production for the development of smart livestock environments. *Internet of Things : Engineering Cyber Physical Human Systems*, *22*, 100724. DOI: 10.1016/j.iot.2023.100724

Morchid, A., El Alami, R., Raezah, A. A., & Sabbar, Y. (2023). Applications of internet of things (IoT) and sensors technology to increase food security and agricultural Sustainability: Benefits and challenges. *Ain Shams Engineering Journal*, •••, 102509.

Pachiappan, K., Anitha, K., Pitchai, R., Sangeetha, S., Satyanarayana, T., & Boopathi, S. (2024). Intelligent Machines, IoT, and AI in Revolutionizing Agriculture for Water Processing. In *Handbook of Research on AI and ML for Intelligent Machines and Systems* (pp. 374–399). IGI Global.

Ponnusamy, V., & Natarajan, S. (2021). Precision agriculture using advanced technology of IoT, unmanned aerial vehicle, augmented reality, and machine learning. *Smart Sensors for Industrial Internet of Things: Challenges, Solutions and Applications*, 207–229.

Raj, M., Gupta, S., Chamola, V., Elhence, A., Garg, T., Atiquzzaman, M., & Niyato, D. (2021). A survey on the role of Internet of Things for adopting and promoting Agriculture 4.0. *Journal of Network and Computer Applications*, *187*, 103107. DOI: 10.1016/j.jnca.2021.103107

S., B. (2024). Advancements in Optimizing Smart Energy Systems Through Smart Grid Integration, Machine Learning, and IoT. In *Advances in Environmental Engineering and Green Technologies* (pp. 33–61). IGI Global. DOI: 10.4018/979-8-3693-0492-1.ch002

Salam, A. (2024). Internet of things in agricultural innovation and security. In *Internet of Things for Sustainable Community Development: Wireless Communications, Sensing, and Systems* (pp. 71–112). Springer. DOI: 10.1007/978-3-031-62162-8_3

Samikannu, R., Koshariya, A. K., Poornima, E., Ramesh, S., Kumar, A., & Boopathi, S. (2022). Sustainable Development in Modern Aquaponics Cultivation Systems Using IoT Technologies. In *Human Agro-Energy Optimization for Business and Industry* (pp. 105–127). IGI Global.

Sankar, K. M., Booba, B., & Boopathi, S. (2023). Smart Agriculture Irrigation Monitoring System Using Internet of Things. In *Contemporary Developments in Agricultural Cyber-Physical Systems* (pp. 105–121). IGI Global. DOI: 10.4018/978-1-6684-7879-0.ch006

Venkateswaran, N., Kiran Kumar, K., Maheswari, K., Kumar Reddy, R. V., & Boopathi, S. (2024). Optimizing IoT Data Aggregation: Hybrid Firefly-Artificial Bee Colony Algorithm for Enhanced Efficiency in Agriculture. *AGRIS On-Line Papers in Economics and Informatics, 16*(1), 117–130. DOI: 10.7160/aol.2024.160110

Vennila, T., Karuna, M., Srivastava, B. K., Venugopal, J., Surakasi, R., & Sampath, B. (2022). New Strategies in Treatment and Enzymatic Processes: Ethanol Production From Sugarcane Bagasse. In *Human Agro-Energy Optimization for Business and Industry* (pp. 219–240). IGI Global.

Xu, J., Gu, B., & Tian, G. (2022). Review of agricultural IoT technology. *Artificial Intelligence in Agriculture, 6*, 10–22. DOI: 10.1016/j.aiia.2022.01.001

KEY TERMS AND DEFINITIONS

AI: Artificial Intelligence.

API: Application Programming Interface.

BI: Business Intelligence.

CO: Carbon Monoxide.

DSS: Decision Support System.

GIS: Geographic Information System.

GPS: Global Positioning System.

HVAC: Heating, Ventilation, and Air Conditioning.

IBM: International Business Machines Corporation.

IoT: Internet of Things.

JSON: JavaScript Object Notation.

LED: Light Emitting Diode.

LPWAN: Low Power Wide Area Network.

ML: Machine Learning.

PII: Personally Identifiable Information.

RFID: Radio Frequency Identification.

ROI: Return on Investment.

SDI: Spatial Data Infrastructure.

VRI: Variable Rate Irrigation.

WAN: Wide Area Network.

XML: eXtensible Markup Language.

Chapter 9
AI-Enhanced Advanced Aquaponics in Agricultural Systems

T. A. Mohanaprakash

Department of Computer Science and Engineering, SOET, CMR University, Bengaluru, India

A. Bhagyalakshmi

Department of Computer Science and Engineering, Vel Tech Rangarajan Dr. Sagunthala R&D Institute of Science and Technology, Chennai, India

S. Senthilkumar

https://orcid.org/0000-0001-8207-2515

Department of Computer Science and Engineering, Vinayaka Mission's Kirupananda Variyar Engineering College, Vinayaka Mission's Research Foundation, Salem, India

Pramoda Patro

https://orcid.org/0000-0003-3942-3927

School of Computer Science and Artificial Intelligence, SR University, Warangal, India

Nynalasetti Kondala Kameswara Rao

Department of Computer Science and Engineering, Sagi Rama Krishnam Raju Engineering, Bhimavaram, India

Sampath Boopathi

https://orcid.org/0000-0002-2065-6539

Department of Mechanical Engineering, Muthayammal Engineering College, Namakkal, India

DOI: 10.4018/979-8-3693-2320-5.ch009

ABSTRACT

The chapter focuses on artificial intelligence in its integration into modern systems of aquaponics and on its modernizing potential in contemporary agriculture. AI-driven solutions enhance the efficiency, sustainability, and productivity of aquaponics through optimization of water quality, nutrient cycling, and fish-plant interactions. This chapter will look into the applications of AI in machine learning algorithms for predictive analytics, automated monitoring systems for real-time data collection, and decision support systems in dynamic resource management. It also involves scaling the operations of aquaponics through the role AI can play in increasing crop yield while reducing the impact on the environment. Case studies have been done showing successful implementations of AI-enhanced aquaponics, showing system resilience and efficiency in resource use. Finally, the chapter concludes with the indication of future directions for AI in aquaponics as a pathway toward more sustainable and intelligent agricultural practices with environmental sustainability goals.

INTRODUCTION

Basically, it is a step ahead in sustainable agriculture, since it is the integration of aquaculture—the growing of fish—with hydroponics, or the growing of plants in water. The way it works is the symbiotic system where fish waste provides essential nutrients for the plants, and the plants filter and purify the water and then return it to the fish tanks. Closed-loop, it wastes very few resources and offers a more environmentally friendly alternative to conventional farming methods. Although it offers many advantages, the very nature of an aquaponic system is complex: It requires the management of biological, chemical, and environmental variables that ensure good health and productivity of both fish and plants (Bali & Singh, 2024).

Artificial Intelligence has been an innovative way of doing agriculture in the past years. Artificial Intelligence offers very effective tools to model, optimize, and eventually automate different processes involved in aquaponics. The latter would therefore enhance efficiency, reduce labour and improve system resilience. With increasing global pressure on the production of sustainable food, AI enhanced aquaponics systems have emerged as potential contributors to food security especially in resource-constrained and environmentally stressed regions (Chandramenon et al., 2024).

In part, the adoption of AI in aquaponics is a part of the new wave in precision agriculture, whereby technology is mobilized to support data-driven farming practices that are more efficient and more sustainable. In this regard, there has been an increase in the application of technologies such as machine learning, computer vision, and predictive analytics in monitoring and control of aquaponic systems. These technologies realize real-time data collection and analysis to achieve the precise adjustment of water quality, nutrient levels, and environmental conditions. For example, AI can predict when levels of certain nutrients are likely to fall below optimal thresholds, which might enable the triggering of automated nutrient dosing (Liu & Jiang, 2024). Similarly, AI-driven computer vision systems could be used to monitor plants' health and growth by noticing signs of stress or sickness that may need intervention.

One of the advantages AI offers to aquaponics is processing large amounts of data and finding trends or patterns within them that may elude human operators. This can be applied within an average Aquaponics system where variables of pH, temperature, Dissolved Oxygen, and nutrient concentrations are continuously monitored to maintain their optimal levels. Doing these manually is time-consuming and prone to errors, especially for larger operations. On the other hand, artificial intelligence systems can process data from multiple sensors simultaneously for a more holistic view of the health of the system and make recommendations or adjustments in real time. This improves the efficiency of the whole system while reducing the potential for human error, culminating in results that are not only consistent but also reliable for production (Choi et al., 2023).

Moreover, there is immense potential for scalability in AI-enhanced aquaponics systems. In most cases, conventional systems in aquaponics are limited because they require constant monitoring and manual intervention. This becomes impractical with the rise in the size of operations. Since AI automates key processes and provides real-time insight, it enables the scaling of aquaponics systems to meet larger production demand without the system integrity or resource efficiency being interfered with. Of particular note, scalability becomes most critical in the arena of urban agriculture, where space and resources are limited and there is a growing need for more sustainable high-yield farming solutions (Channa et al., 2024).

Not only can AI enhance operational efficiency and scalability, but it can also help improve the sustainability of aquaponics systems. One of the great environmental benefits of aquaponics is its reduced usage of water as compared to traditional, soil-based agriculture. This efficiency is predicated on management that ensures optimum water quality and efficient cycling of nutrients, both of which can be optimized with AI. By providing very precise control over the inputs and continual monitoring of the outputs, AI can aid in reducing waste and ensure that the system functions within sustainable limits. Moreover, AI will help in developing closed-loop systems that can recycle and reutilize water and nutrients more efficiently, further reducing the environmental footprint from aquaponics (Karimanzira & Rauschenbach, 2021).

While to some extent aquaponics can be driven by AI, it also does not come with few challenges. First, there have to be upfront technological and expertise investments for the development and implementation of AI systems. Besides, robust data management practices are required in order to make sure that AI-driven insights are not only accurate but also reliable. Another concern is that of data privacy and security in large-scale operations where sensitive information may be at stake. Even with these challenges, the potential benefits from AI-enhanced aquaponics systems are huge in setting them on the right course toward sustainable, resilient, and productive agriculture (Jiang & Liu, 2024).

Applications of AI in aquaponics will grow as the field does. That could entail further development of more sophisticated AI models predicting and reacting to complex interactions within the system in the future and integrating AI with other developing technologies such as the Internet of Things and blockchain. It is still possible that such innovative steps further ahead may better the efficiency, transparency, and traceability of aquaponics, making this sector more attractive to producers and consumers alike (Rozie et al., 2020).

Finally, integration of artificial intelligence in advanced systems marks another frontier for sustainable agriculture. Aquaponics operations will be more efficient and scaled up in a way that is environmentally friendly through the use and optimization of AI's capabilities in data analysis, automation, and decision support. The fact that climate change, resource scarcity, and population growth are likely to prevail as push factors for innovative solutions in agriculture means that AI-enhanced aquaponics could very easily assume a critical role in meeting food production needs of the world while minimizing environmental impact (Abbasi et al., 2022). This chapter reviews different AI technologies being applied in aquaponics, their benefits, and challenges; future directions in what is sure to be an exciting and rapidly evolving field are also considered.

Background

Probably one of the most novel agricultural practices is aquaponics—a symbiotic environment combining aquaculture, which is fish farming, and hydroponics, meaning the cultivation of plants without using soil. Traditionally, aquaponics has been hailed as an efficient and water- and nutrient-saving approach; therefore, standing as a viable alternative to conventional methods of farming. In contrast, managing these complex interactions between aquatic and plant systems runs into a number of issues related to maintaining appropriate water quality, balance of nutrient levels, and health for both the fish and the plants. With technological advancement, AI comes forward as an able tool in transcending these challenges. AI can automate monitoring processes, make predictions of system behaviors, and optimize resource use, therefore making aquaponics more efficient and scalable. The chapter capitalizes on the growing interest in AI's role in agriculture, particularly with regard to aquaponics, which aspires to explore how such technologies can be utilized toward revolutionizing this field for increased sustainability, productivity, and resilience of agricultural systems.

Scope

The scope of this chapter is AI applications in cutting-edge aquaponics systems, where theoretical dimensions are paid attention to along with practical ones. In this chapter, some AI technologies—in particular, machine learning algorithms, predictive analytics, and automated monitoring systems—are considered against the background of how they help optimize aquaponics operations. Critical aspects are covered, including water quality control, nutrient management, and dynamic interactions between aquatic and plant life. The chapter also covers the investigation of scalability in AI-enhanced aquaponics, looking into challenges and opportunities such systems will present at different operational scales. Case studies were used to derive insights as to what AI is really doing for aquaponics in practice, focusing on sustainability, efficiency in resources mobilization, and crop yield enhancement. This chapter assumes to some extent that AI will give an overview of this area concerning the role of AI in the field of agriculture, to help enlighten researchers, professionals, and policymakers about its future promise and way forward for sustainability.

Objectives

This chapter primarily is geared toward the integration of AI technologies into advanced aquaponics systems aiming for increased efficiency, sustainability, and scalability. It tries to take a view of how AI-driven solutions, such as machine learning and automated monitoring, could optimize critical processes like water quality management, nutrient cycling, or fish-plant interactions. The chapter also aims to make available some case studies that clearly demonstrate the practical benefits of AI in improving crop yield, reducing environmental impact, and supporting resource-efficient practices. Finally, this chapter also aims to draw out lessons for the future of AI-enhanced sustainable agriculture.

FUNDAMENTALS OF AQUAPONICS

Aquaponics is a sustainable agriculture method where, in a closed-loop cycle, it merges aquaculture. This interaction helps the fish and plants grow from each other's biology, thus attaining an efficiently functioning and highly eco-friendly farming method compared to traditional means. To fully understand the extent of aquaponics, there needs to be an exploration of its historical development, its key components, and how the symbiotic relationship is channeled into success (Styawati et al., 2023).

Historical Development of Aquaponics

The concept of aquaponics dates back thousands of years to ancient cultures from around the globe. The ancient Aztecs used to grow plants on rafts in shallow lakes, a technique known as "chinampas," which was a form of space-saving aquaponics in action. Plants grown on floating gardens would be able to directly absorb the waterborne nutrients into the roots (Boopathi et al., 2023; P. Kumar et al., 2023; Samikannu et al., 2022).

Similarly, rice paddies in Asia have long used a system whereby fish were cultured in the flooded fields, improving soil fertility while providing some form of pest control and receiving an added source of food. Although all these systems were much less technologically intensive than today's aquaponics, they laid the basis for these integrated farming practices today.

Modern aquaponics, as we know it, began to take shape in the 20th century, with the most prominent developments happening in the 1970s. It was during this period that researchers and agricultural innovators started experimenting with the combination of aquaculture and hydroponics within controlled environments. These early experiments were interested in understanding the nutrient dynamics between fish and plants and in finding out ways through which the needs of the two components could be balanced. Due to this, aquaponics systems evolved further to increase their productivity and give way to the most diverse and scalable systems we can see today.

Important Components: Aquaculture and Hydroponics

An aquaponics system includes two interlinked major subsystems: aquaculture and hydroponics. Fully understanding each component is important to appreciate how they work together to create a balanced, sustainable system (Taha et al., 2022).

In the context of aquaponics, aquaculture refers to rearing aquatic organisms, mostly fish, in an Artificial environment. This will have the fish being grown in tanks where their excretory products, mainly ammonia, are done. Were this to be an aquaculture-only system, then this would have to be removed and treated for toxic buildup, which could prove quite harmful to the fish. In an aquaponics system, though, that waste becomes quite valuable. The bacteria first break down the ammonia into nitrites, then into nitrates. In turn, the nitrates become very important plant nutrients. Tilapia, catfish, and trout are popularly farmed fish in aquaponics due to their strong resistance and ability to grow very fast.

Hydroponics refers to a method through which plants are grown in a nutrient-rich water solution without the use of soil. During the process of aquaponics, the nutrient solution is derived from the excretion of fish. The excreted material is then converted by bacteria into forms absorbable by plants. In this system, plants are usually grown in a soilless medium, such as gravel, clay pellets, or coconut coir, which gives support while allowing roots access to the water rich in nutrients. These plants, therefore, grow on these same nutrients but also form a critical function of filtering and cleaning the water before it is recirculated back to the fish tanks.

Symbiosis in Aquaponics Systems

This kind of interrelation between fish and plants is the root of aquaponics systems. This mutually advantageous interaction makes aquaponics very different from other ways to grow produce and distinctly sustainable. Fish in an aquaponics system excrete waste into the water, mainly in the form of ammonia. While ammonia is usually toxic to fish if its concentration becomes high, for plants it serves as one of the most important sources of nutrition. These nitrifying bacteria naturally present in the system convert ammonia first to nitrites, then to nitrates, readily available to plants. During this process of conversion, the water gets detoxified not only for the fish but is also enriched with the proper nutrients for the robust growth of plants (Hui et al., 2022).

These, in turn, are absorbed by plants and other plant foods, cleaning the water as it passes through the hydroponic part. The water, now cleaned, is then circulated back into the fish tanks, closing the loop. In this way, this continuous resource exchange is very nearly self-sufficient; each by-product from one element is the supply for another, greatly eliminating the reliance on a large number of possible chemical fertilizers or water treatments inputted into the system.

More than that, the efficiency of aquaponics is in terms of the maximization of resource use while minimizing waste. Water is conserved through re-circulation, and reliance on chemical fertilizers is drastically reduced since plants obtain their nutrients directly from fish waste. A closed-loop system, it is not only environmentally friendly but cost-effective, thus very appealing for producing sustainable food.

Aquaponics represents a convergence of knowledge evolved through traditional agriculture and today's technological leaps, forming a highly efficient and sustainable way of growing food. Understanding its process of historical development, the roles of aquaculture and hydroponics, and symbiotic relationships that drive these systems give insight into how aquaponics can be used to answer global challenges related to food security, resource scarcity, and environmental sustainability. In the new wave of interest in sustainable agriculture, aquaponics has emerged as a leading solution for food production that is harmonious with natural ecosystems.

AI TECHNOLOGIES IN AQUAPONICS

Artificial Intelligence can change the face of aquaponics by bringing improvement in system efficiency, optimizing resource use, and better crop yields. AI embeds advanced technological capabilities into an aquaponics system with improved monitoring, prediction, and control over a given parameter. The purpose of this section is to explore how various

AI technologies—Machine Learning and Predictive Analytics, Computer Vision for Plant and Fish Health Monitoring, Automated Monitoring and Control Systems, Decision Support Systems—are shifting aquaponics onto a more sustainable and scalable farming practice (Hui et al., 2022; Metin et al., 2023).

Figure 1. Technologies in aquaponics

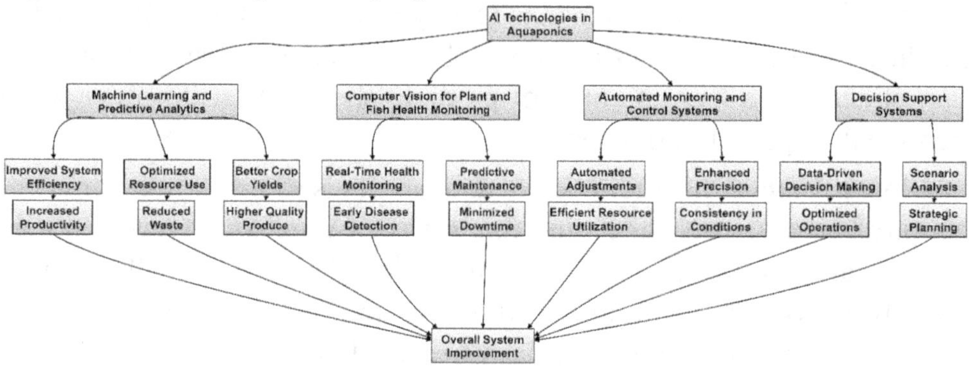

Figure 1 illustrates the AI technologies contribute to enhancing aquaponics systems, from improving efficiency and optimizing resource use to enhancing crop yields.

Machine Learning and Predictive Analytics

ML is probably at the forefront of AI applications in aquaponics. This ability to analyze huge amounts of data and generate predictive models that can predict system behavior makes ML probably at the forefront of AI applications in aquaponics. An aquaponics setup calls for continuous monitoring and balancing of a lot of variables, such as water temperature, pH levels, concentration of nutrients, and fish growth rates. Such parameters can be used by algorithms to analyze historical data for patterns and trends that would not be easily recognizable by immediate observation (Glady et al., 2024; Pachiappan et al., 2024).

For example, ML can predict the most opportune feeding schedules for fish to be given the right quantity of food at the right time. This reduces wastage while enhancing growth rates. Predictive analytics also help in foreseeing plant nutrient needs, whereby the correct adjustments in the delivery of nutrients ensure optimal plant health and yield. Such predictions from ML models can ensure preemptive maintenance or adjustment of systems that are prone to failures or inefficiencies, reducing the time of no production and preventing crop loss.

Computer Vision for Plant and Fish Health Monitoring

AI, specifically the domain of computer vision, enables machines to understand and act on visual data. Therefore, it is very instrumental in monitoring the health of both plants and fish in aquaponics systems. High-resolution cameras and sensors deployed across the system take real-time images and videos of the plants and fish. Using AI algorithms, these images are then analyzed to detect the early signs of a disease, stress, and suboptimal growing conditions of plants and fish.

Computer Vision in plant health monitoring can ascertain whether the plant has some kind of problem due to deficiency, pests, or fungal attack by its color, shape, and leaf texture. It can also, in fish health monitoring, identify abnormalities in fish behaviors—for instance, erratic swimming patterns, which would point toward disease or poor water quality. It enhances the possibility of timely interventions through the constant monitoring and analysis of health indicators, hence improving overall system productivity and reducing losses due to disease or poor health.

Automated Monitoring and Control Systems

The delicate balance needed to be maintained in aquaponics relies on AI-powered automated monitoring and control systems. These systems have used, and continue to use, sensors and IoT devices to trace key environmental parameters continuously: water temperature, pH levels, dissolved oxygen, nutrient concentrations. The data from these sensors gets fed into an AI-driven control system that enables it to make adjustments to the conditions automatically and maintain optimum levels.

For example, if the water temperature is out of range from the ideal for fish growth, then the AI system can trigger heating or cooling mechanisms to bring back appropriate temperatures. Similarly, if the pH level is too acidic or alkaline, the system will be able to auto-dispense the required chemicals to neutralize the water. The automated adjustments ensure that the aquaponics system remains stable and productive without the need for constant human oversight.

Besides, it can optimize the usage of energy and resources to make aquaponics more sustainable. For example, AI can control lighting and water circulation systems according to real-time needs, thus potentially saving electricity and water.

Decision Support Systems in Aquaponics

DSSs for aquaponics are much more than management tools; they integrate artificial intelligence in providing the farmer or operator with appropriate insights to enhance the decision-making process. They derive from a large diversity of data, such as environmental sensors, historical records, and external weather forecasts, in order to present a detailed guide on managing these systems (P. Kumar et al., 2023; Kumara et al., 2023; Zekrifa et al., 2023).

This tool could be used in scheduling crop rotation, choosing the right fish species, and deciding on the harvest time to give maximum yields. By considering the data available and past records, it will be easy for DSS to give a general overview of how to handle the system effectively by dealing with various seasons and other changes in the environment.

Additionally, DSS can integrate economic variables by making optimum cost-saving and profit-maximizing decisions available to operators. It could advise on changing feeding regimens with respect to the prevailing fish feed market prices without altering the required nutritional values; that is, cost-effectiveness remains intact. Other than that, DSS can project the probable market demand for a specific crop to be grown and thus let the farmer alter production strategy accordingly.

It is in this regard that AI technologies are being integrated into aquaponics as one huge leap forward in sustainable agriculture. Machine learning, predictive analytics, computer vision, automated monitoring and control systems, and decision support systems can achieve higher efficiency, productivity, and sustainability in aquaponics. These AI-driven technologies enhance operational capabilities of aquaponics systems and further bring forward new opportunities for innovation in farming: producing food more sustainably and at scale, even in challenging environments. Continuing the development path, it is without a doubt that AI will increase its applications in aquaponics and continue to change the way food production and resource management are approached.

OPTIMIZATION OF AQUAPONICS SYSTEMS THROUGH AI

AI in aquaponics systems deals with better efficiency and effectiveness of different components to make sure that water quality, nutrient balance, and environment conditions are managed in the best possible way. AI technologies make it possible to monitor the system on a real-time basis and maintain predictions for correcting the system in the best possible fashion, which in turn improves system performance and sustainability. The next section discusses the use of AI in aquaponics systems optimization with regard to water quality management, nutrient cycling and balance, monitoring of environmental conditions, analysis of real-time data, and system adjustment (P. Kumar et al., 2023).

Figure 2. AI in optimizing aquaponics systems

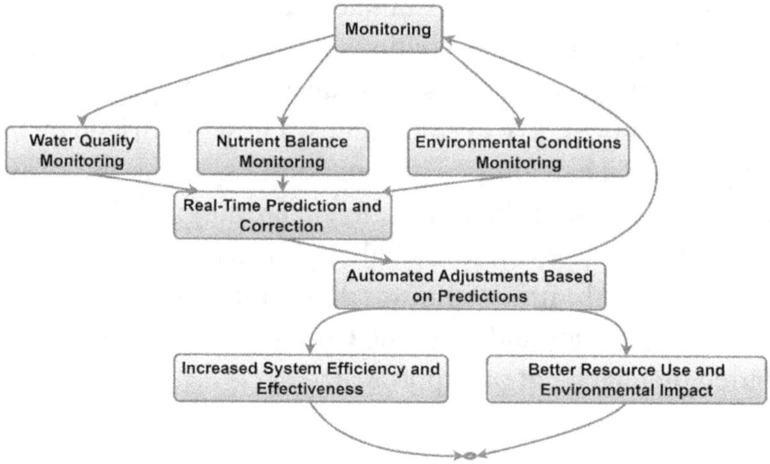

The diagram in Figure 2 showcases the use of AI in optimizing aquaponics systems, focusing on water quality, nutrient balance, and environmental conditions, thereby enhancing system performance and sustainability.

Water Quality Management

An optimum water quality is essential for sustaining the health of fish and plants in an aquaponics system. AI regulates the water quality of an aquaponics system through continuous monitoring and observation of parameters such as pH, temperature, dissolved oxygen, and ammonia levels (Samikannu et al., 2022).

Real-Time Monitoring: AI-driven sensors and IoT devices continuously measure the water quality parameters in this. The data from these sensors is continuously fed into AI algorithms, which analyze them against any deviation from the optimum range. For instance, if there is a rise in ammonia levels on account of fish wastes, AI systems can promptly identify the increase and trigger corrective actions such as adjusting filtration rates or increasing aeration.

Predictive Analytics: Machine learning models can be trained to predict water quality problems before they actually reach a critical level. Scanning the past record and detecting trends, Artificial Intelligence can make forecasts of when water quality might fall below certain standards given fish biomass and feeding rates. Such an early warning system would enable timely interventions to replace water or adjust filtration systems so that the environment remains friendly for fish and plants to thrive in.

Automated Control Systems: Artificial intelligence merges with automated control systems, thus changing the parameters of water quality dynamically. For example, if it reads a too acidic pH, it will be able to automatically add a buffering agent to re-balance it. By the same token, AI could hold the temperature and oxygenation to required values from the real-time data through systems, keeping conditions steady and in an ideal range for fish and plants.

Nutrient Cycling and Balance

Nutrient cycling and balancing are important in maximizing plant growth but also in ensuring fish receive sufficient nutrition, and AI supports this through optimization of its distribution and balance in the system (Boopathi et al., 2023).

Nutrient monitoring: The AI-powered sensors monitor the levels of nitrogen, phosphorus, and potassium, major plant nutrients, in the water. The sensors continuously respond to the concentration of these nutrients, and AI systems adjust the rate of delivery of those nutrients appropriately. Suppose the nitrogen level is down from its optimal window; in this instance, the supply of compounds rich in nitrogen can be raised.

AI models consider data on plant nutrient intake and fish waste production to optimize nutrient cycling. Such models could easily predict, given plant growth stages and fish feeding rates, nutrient requirements that would keep the system within an appropriate nutrient balance. In effect, AI predicts demand in advance, thus helping to avoid deficiencies or excesses, which may lead to reduced plant health or fish stress.

Dynamic adjustment: With the availability of real-time data, it is easy to adjust the concentration of nutrients in the plant. For instance, if there is an indication that a certain plant does not have a particular nutrient, then the AI will adjust the solution composition to suit the plants' needs.

This focused effort at improving the health and maximizing growth of the plants leads to a balanced environment for fish.

Environmental Condition Monitoring

The success of any aquaponics setup largely depends on monitoring and managing environmental conditions. AI helps in elaborately monitoring environmental conditions, gaining automated control over the applied lighting, humidity, and the level of CO_2 (Vangeri et al., 2024; Vijaya Lakshmi et al., 2024).

Comprehensive Sensing: AI fuses data from different sensors to monitor the environment of an aquaponics system. For instance, a microclimate characterized by plants and fish is monitored for temperature and humidity, as well as CO_2 levels; these environments are read from this data to assure that conditions are optimal for both plants and fish.

Predictive Adjustments: AI uses historical data and predictive models to project the changes in the environment and their impact on the system. For example, AI may predict temperature fluctuations based on changes in the seasons and, respectively, it will adjust heating or cooling systems. Thus, by predicting changes, AI sustains a stable environment with less stress on plants and fish.

Automated Controls: Using AI-based environmental sensors, the conditions of the environment are controlled automatically. An instance of this is that, when there is a realization that sensed light conditions are less than conducive for the growth of plants, then AI will adjust the extent and period settings of artificial lights accordingly. Equally, AI systems regulate ventilation and humidity levels to ensure a constant and healthy environment.

Real-Time Data Analysis and System Adjustments

Real-time data analytics lies at the very essence of any aquaponics system optimization, enabling real-time responses to changing conditions while keeping up the system's efficiency(P. Kumar et al., 2023).

- **Data Integration:** AI systems integrate data from such diverse sources as environmental sensors, fish health monitors, and plant growth trackers. With such a holistic view into your data, AI is better positioned to analyze these complex interactions for informed decisions on system adjustments.
- **Adaptive Algorithms:** Machine learning algorithms are adaptive because of the changing conditions and hence will learn from ongoing data. For example, it is possible for AI to fine-tune models based on feedback in real-time to improve the precision of its predictions in managing the system parameters. By this, the system stays responsive against any expected or unexpected changes.
- **Automated Responses:** Artificial intelligence now enables automated responses to real-time data and is timely in terms of interventions. For instance, if a sensor picks up an anomaly in water quality, then AI systems can automatically trigger the correction, such as adjusting filtration or nutrient levels. This automation reduces manual intervention and enhances system stability.
- **Continuous Improvement:** AI-driven analytics provide insights that support continuous improvement. Value stream analysis on performance data could help in the identification of areas of optimization, propose enhancements to system design, or even recommend certain changes in management practices. In the long run, these unending optimizations will lead to success and sustainability in an aquaponics system.

AI technologies in aquaponics system optimization are moving fast in changing water quality management, nutrient cycling, environmental condition monitoring, and real-time data analysis. Integrating such technologies into an aquaponics system will ensure increased efficiency, better utilization of resources, and increased sustainability. Application of AI in aquaponics does not only enhance operational capabilities but also fosters the creation of more resilient and productive agricultural systems to help feed humanity and ensure environmental sustainability.

SCALABILITY AND EFFICIENCY IN AI-ENHANCED AQUAPONICS

Artificial intelligence integrated into the aquaponics systems makes very promising improvements in terms of scalability and efficiency, breaking thus the barriers of conventional techniques to extend possibilities for sustainable agriculture. The next section describes issues connected with the scaling of traditional aquaponics and the role of AI in the empowerment of large-scale operations, with special attention paid to AI applications in urban agriculture and vertical farming(Koshariya, Khatoon, et al., 2023).

Figure 3. Scalability and efficiency in AI-enhanced aquaponics

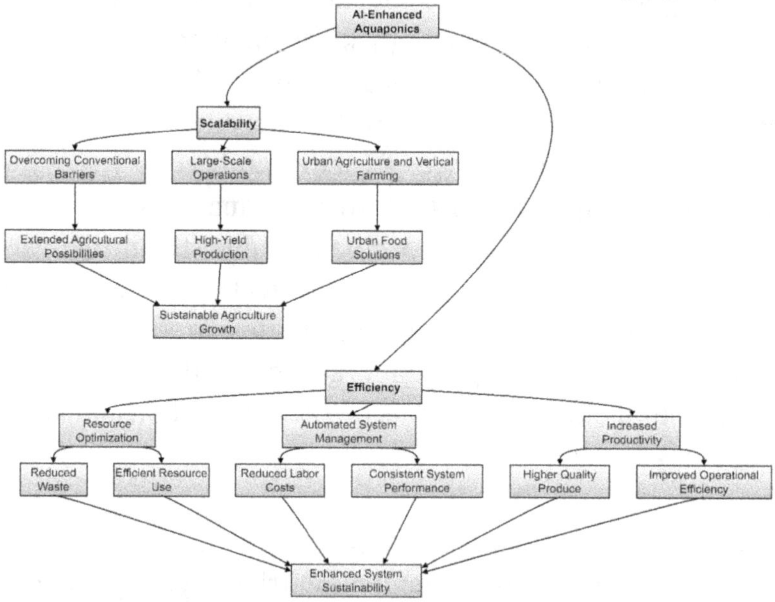

Figure 3 showcases the integration of AI into aquaponics systems, enhancing its scalability and efficiency.

Challenges of Scaling Traditional Aquaponics

There are many issues when scaling traditional aquaponics systems because of the originally complex design and related resource demands. Some of these include(Gnanaprakasam et al., 2023; Koshariya, Kalaiyarasi, et al., 2023):

System Complexity: Conventional aquaponics systems function on an intricate balance between aquaculture and hydroponics. In scaling, this becomes progressively more complex. Large systems need more sophisticated mechanisms for monitoring and controlling water quality, nutrient levels, fish, and plant health.

Resource Intensiveness: At any rate, increasing the size of an aquaponics system requires, most of the time, huge investments in infrastructure, entailing large fish tanks, grow beds, and filtration systems. This may represent high costs and a heavy log when scaling up, especially for smaller- to medium-sized companies.

Harmonization: Labor and Management As an aquaponics system becomes larger, so do the monitoring and maintenance needs. The size of the system makes the continuity of labor and timeliness of interventions harder to attend to—an inefficiency that eventually leads to escalating costs of operation within an aquaponics system.

Environmental and Space Constraints: Traditional ways of aquaponics are mostly area-consuming and resource-intensive, not viable in urban or high-density population areas. Scaling up such systems while managing the constraints of space and environmental pressures is extremely difficult.

Role of AI in Large-Scale Operations

AI works magic in playing a transformative role in the tending of challenges in scaling aquaponics systems with increased efficiency and feasibility of large-scale operations (Agrawal, Magulur, et al., 2023). Contributions of AI majorly involve:

Advanced Automation and Control: Intelligence-driven automation systems autonomously drive large-scale aquaponics with ease in terms of routine tasks, such as monitoring water quality, adjusting the levels of nutrients, and regulating environmental conditions. It reduces manual intervention and enables operators to more efficiently and effectively manage larger systems.

Predictive Analytics and Optimization: AI models apply predictive analytics in predicting the performance of the system to detect problems earlier before their situation worsens. Through analysis of historic and real-time data, AI can optimize feeding schedules, nutrient delivery, and environmental controls to ensure smooth running and resource efficiency in large-scale operations.

Scalability of Data Management: Artificially Intelligent systems can process and analyze such vast volumes of data created by large-scale aquaponics operations. Machine learning algorithms process data from numerous sensors and sources and turn it into actionable insights that drive data-driven decision-making. This functionality becomes very important when one is dealing with complex and large-scale systems where traditional methods might be incapable of keeping up.

Resource Efficiency: Through AI, resource efficiency is supported with minimized water usage, energy, and application of nutrients. For instance, on a real-time basis, AI systems automatically adjust the re-circulation rate of water and lighting. It reduces wastage, hence allowing for a reduction in operational costs. This efficiency is more critical in large-scale operations where resource management can make quite a difference in overall sustainability.

Urban Agriculture and Vertical Farming Applications

AI-enhanced aquaponics is very well suited for urban agriculture and vertical farming (Agrawal, Shashibhushan, et al., 2023; P. R. Kumar et al., 2023). These are areas with special challenges, such as:

Optimization of Urban Spaces: Space is a limitation in an urban setup, and it may not assist in setting up the traditional aquaponics systems. Artificial intelligence facilitates the construction of compact and efficient systems that optimize available spaces to the fullest. One of the most promising applications for artificial intelligence in agriculture seems to be vertical farming, also known as a production method wherein plants are grown either in stacked layers or on vertically inclined surfaces. The AI-driven optimization of light, water, and nutrient distribution supports high-density food production within confined spaces.

Smart City Integration: AI-enhanced aquaponics could be applied within smart city frameworks, contributing to the goals of urban sustainability. AI can manage aquaponics installations as a part of broader smart infrastructure in ways that optimize energy use, reduce waste, and provide food security locally. All this ties back to urban initiatives aiming at resilient, self-sufficient communities.

Efficient Use of Resources: Most of the time, urban agriculture suffers from insufficient water and energy. AI helps overcome such limitations by optimizing resource use within aquaponics systems. For example, AI can control hydroponic nutrient solutions and water recycling processes to make sure resources are used efficiently to avoid waste and really help in sustainable urban food production.

Scalability in Urbanized Areas: AI will allow these systems to scale up in urbanized areas by enabling modular designs and remote manageability of production. Scalable in small units allows the producer to increase the number of units during demand increase in a step-by-step manner. AI-based remote monitoring and control systems allow operators to oversee multiple Urban Aquaponics installations from a single location, hence increasing their overall scalability and operational efficiency.

Fully integrated AI in aquaponics systems resolves the scaling issues of conventional methods, improving efficiency and feasibility in large-scale operations. AI technologies create huge benefits in regard to automation, predictive analytics, data management, and optimization of resources that enable the scaling of aquaponics systems more effectively. Additionally, in urban agriculture and vertical farming, AI makes optimal use of available space, integrates into smart city initiatives, and results in efficient resource usage. As AI continues to explore new frontiers, its potential role in aquaponics transformation seems to further increase when it comes to increasing sustainability and scalability of agricultural solutions in diversified environments.

SUSTAINABILITY AND ENVIRONMENTAL IMPACT

AI-enhanced aquaponics systems open up a path towards a more sustainable and environmentally friendly agriculture, aimed at optimizing the use of resources, minimizing wastages in the cycle, and integrating closed-loop advanced systems. This section shall look into how AI aids in sustainability and the environment through lesser use of resources, closed-loop systems, and case studies of successful AI-enhanced aquaponics practice (Glady et al., 2024; Venkateswaran et al., 2024).

Figure 4. Sustainability and environmental impact of AI-enhanced aquaponics systems

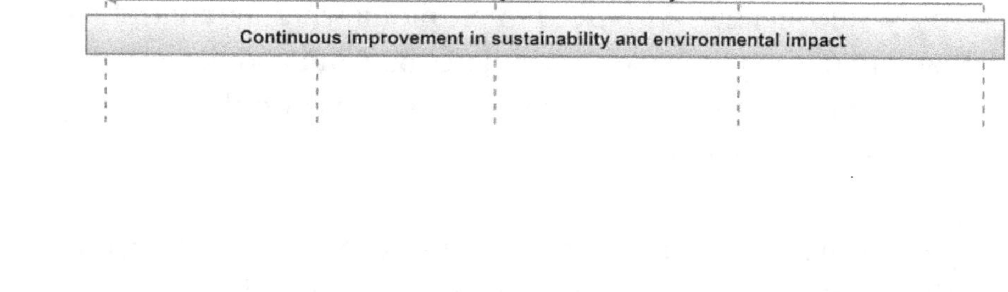

AI-Reduced Use of Resources

Artificial intelligence significantly reduces the use of resources within an aquaponics system by managing key parameters such as water, energy, and nutrients.

- **Water Efficiency:** AI technologies enhance the management of water usage by monitoring it and regulating it properly. Advanced sensors and machine learning algorithms provide real-time monitoring of both water quality and water usage patterns. According to growth stages of plants and fish, AI systems can regulate the water flow rate, optimize recycling processes, and forecast water needs to make efficient use of water without wasting it and thereby reducing frequent water changes.
- **Energy Optimization:** AI enhances energy efficiency through environmental management of light, heating, and cooling. AI-driven systems will evaluate the amount of time required for lighting and its intensity, considering the changing growing demands of plants and available natural illumination. In heating and cooling, predictive models developed by AI will accurately forecast changes in temperature and prompt relevant systems to adjust in time, thereby saving energy and enhancing efficiency.
- **Nutrient Management:** AI algorithms ingested with the current readings of plants and soil with regard to their nutrient levels and compute the actual requirement to supply the needed amount. AI's proactive prediction of nutrient demand will help adjust the nutrient solutions and thereby avoid over-nutrition, which often leads to wastage and environmental pollution. This precision will bring betterment in plant health and reduce the influence on the environment from overflow of these nutrients.

Closed-Loop Systems and Waste Minimization

Closed-loop systems are intrinsic to the sustainability of aquaponics; AI makes these more effective through optimization for waste recycling and minimization of environmental impacts.

- **Nutrient Recycling:** This is an AI-driven closed loop for cycling of nutrients between fish and plants. Being rich in nutrients, fish wastes become the food of plants through biological filtration. Artificial intelligence monitors continuously the level of nutrients and balances the filtering processes so that they get effectively recycled and consumed. Minimal wastages and less dependency on external nutrient inputs.
- **Waste Management:** Artificial intelligence optimizes the breaking down of organic matter and ensures that the waste is put into work. Automated systems can process fish waste or plant debris, turning it into compost or whatever other form may be used. AI-driven sensors and algorithms ensure the efficiency of waste processing and its compliance with the needs of the system, reducing the environmental footprint.
- **Resource Circularity:** AI can be combined with closed-loop systems of aquaponics to increase the resource circularity of the process. Artificial intelligence will then balance resource input with its corresponding output, making sure that resources like water, nutrients, and energy are continuously cycled back and used.

Case Studies on Sustainable AI-Enhanced Aquaponics

The case studies below show how AI-enhanced aquaponics helps to work on sustainability and reduce impacts on the environment:

- **Urban Aquaponics in Singapore:** AI-driven aquaponics systems have been executed in the urban setting of Singapore where there are acute conditions of space and resource shortages. Such systems use AI to optimize water and nutrient management. The amount of water used, along associated energy spent in its circulation, is drastically reduced. Coupling AI with vertical farming techniques can maintain high-density food production within limited spaces, therefore contributing much toward sustainability in cities and local food security.
- **Sustainable Aquaponics in the Netherlands**: Researchers in the Netherlands used AI to optimize closed-loop aquaponics systems.

This AI-driven system is fairly accurate in managing water quality, nutrient levels, and energy consumption within the process. Of course, huge savings have been made on wasted water and nutrients, states the research facility, though the AI application has turned out to be quite successful in achieving sustainable resource use with less environmental impact.

- **AI-Powered Aquaponics in North America:** A North American startup has developed a large, completely AI-powered aquaponics farm that takes sustainability and efficiency to heart. Their AI controls the environment, nutrient supply, and waste. In this way, the startup was able to reduce resources and reduce wastes drastically, proving AI's potential to make large-scale aquaponics operations sustainable.

AI, therefore, is a very critical element in the development of sustainability and the reduction of environmental impact in aquaponics. Via optimized use of resources, closing the loops, and strategies for waste minimization, AI improves the environmental sustainability for such aquaponics. Case studies show that AI delivers practical benefits in bringing sustainable aquaponics practices from urban settlements to large-scale farms. Applications of the AI technology in aquaponics will only increase with its evolution, further aiding sustainable and green agricultural practices for global efforts at reduced use of resources and impact on the environment.

CHALLENGES AND CONSIDERATIONS

While AI is found to offer a lot of benefits in an aquaponics system, it also comes with various challenges and considerations to be effectively harnessed. The different challenges and considerations include technological and financial barriers, management of data, privacy and security concerns, and human factors and integration into current practice(Boopathi, 2024; Pachiappan et al., 2024).

Figure 5. Challenges and considerations of AI integrated aquaponics system

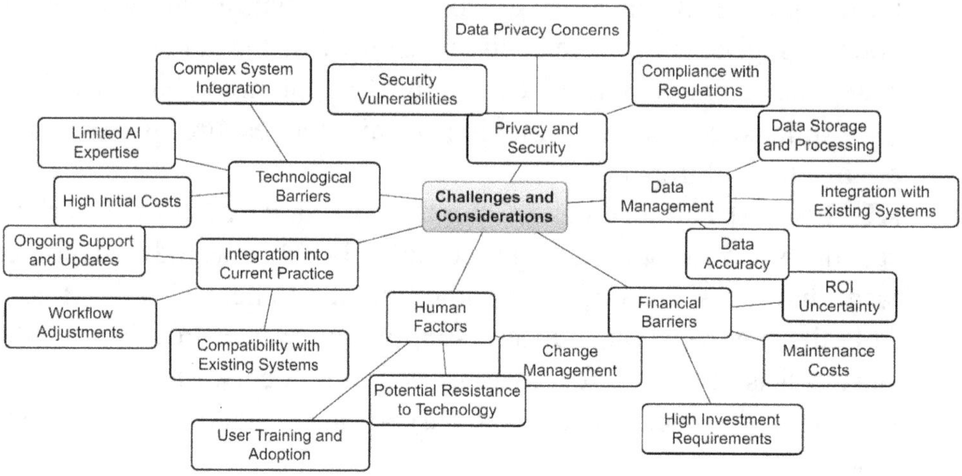

Technological and Financial Barriers

High Initial Investment: Setting up an AI-enhanced aquaponics system requires huge upfront investments. Starting with advanced sensors, AI software, and automation infrastructure, the financial barriers to entry can be quite high for small operators or startups—thereby limiting the access of AI technologies and hindering their diffusion process.

- **Complexity of Technology:** Carrying out the integration of AI into the production of aquaponic systems involves technologies that are very complex and require special kinds of knowledge and competencies in their application. Planning, then maintaining and debugging AI systems demand knowledge in aquaponics and AI technologies. This can result in a high learning curve and technical challenges in such systems, especially to those who lack experience in AI.
- **System Integration:** It is one of the bigger challenges in integrating AI with existing aquaponics systems or infrastructure. Ensuring that new AI technologies are compatible with the older systems—legacy systems—to ensure seamless operation might require custom solutions and additional development, adding cost and possible delays.

- **Maintenance and Upgrades:** The AI systems are in constant need of maintenance and updating to keep them running at their optimum levels. Periodic technical support, software updating, and calibration of the system could raise the long-term cost and complexity of AI-enhanced aquaponics operations.

Data Management, Privacy, and Security

Data Volume and Management: The AI systems generate and process huge amounts of data from sensors and monitoring tools. To manage and analyze such huge data volumes, efficient data management systems and expertise are needed. Dealing with large volumes of data can be very challenging, especially when it comes to being accurate and extracting actionable insights. (Agrawal, Magulur, et al., 2023; Rahamathunnisa et al., 2023; Srinivas et al., 2023)

Privacy: Data from the aquaponics systems could, in some cases, include sensitive information. Examples may be related to the operational details of the system or business performance metrics. Therefore, safeguarding measures should prevent data from unauthorized access by ensuring that it complies with privacy regulations. Subsequently, there should be stringent protection measures of the data to protect any type of proprietary information.

Security Risks: AI systems are open to cybersecurity threats. Any unauthorized access, data breaches, or cyberattacks are bound to break through into the system and cause operational chaos. These can only be guaranteed through adequate security considerations in an AI system, with strong cybersecurity measures and regular security testing in place.

Regulatory Compliance: Businesses that utilize AI technologies must comply with data protection regulations such as the GDPR or CCPA. It could be intricate and complex, often needing legal and technical expertise within the practices of collecting, storing, and processing data to meet the enforced legal obligations. Human Factors and AI Integration

Skill Requirements: Artificial intelligence in an aquaponic setting will require a mix of skills oriented to AI technologies and aquaponics. In view of this, staff training and upskilling to work effectively with AI systems can turn out as the greatest challenge, more so in industries that have few experts in such matters.

Change Management: AI technologies naturally mean changes to the current workflow and processes. Managing this transition requires effective change management strategies that make staff comfortable with new technologies and minimize operational disruptions. It is important to note that resistance to change and problems of adjustment may affect the successful implementation of the AI system.

Human-AI Collaboration: While AI undertakes most of the activities automatically, the supervision of humans is necessary for controlling and interpreting the outputs of AI. Such balancing act in between AI systems and human operators needs to be guided by way of clear communication and collaboration strategies. It is important that the human operator understands AI recommendations and can intervene if needed to maintain system effectiveness and reliability.

Ethical Considerations: As AI will probably take ethical decisions, one has to address the ethical uses of it. However, this is closely related to decision transparency and accountability. The focus has to be in ensuring that AI works in such a manner that is ethical in deciding or taking actions in ways that are valuable for and desirable to human beings for trust and acceptance from users and stakeholders.

The challenges and considerations regarding the implementation and operation of AI-enhanced aquaponics must be balanced, in all related areas: technological, financial, data management, privacy and security concerns, and even human factors. Addressing these challenges will enable the business to work more efficiently, sustainably, and with effectiveness in the operation of aquaponics, contributing to the betterment of agricultural technology and sustainable food production.

Tables 1 and 2 show that AI integration significantly enhances the efficiency and productivity of aquaponics systems, though it comes with higher upfront costs and complexity.

Table 1. Advantages and disadvantages of AI in aquaponics

Aspect	Advantages	Disadvantages
Automation	- Automates monitoring and control of pH, temperature, and nutrient levels. - Reduces the need for manual labor.	- High initial cost for AI system installation. - Dependence on technology may lead to challenges in case of system failures.
Resource Efficiency	- Optimizes water, energy, and nutrient use. - Minimizes waste through precise control.	- Requires regular updates and maintenance of AI systems to stay effective.
Productivity	- Increases yield and consistency of crops. - Improves fish health and growth rates.	- Potentially complex system integration with existing infrastructure.
Data Analysis	- Provides real-time data analytics for better decision-making. - Predictive analytics to prevent issues.	- Data privacy concerns and potential for cyber threats.
Scalability	- Easily scalable for large-scale operations. - Supports vertical farming practices efficiently.	- Requires skilled personnel to manage and interpret AI outputs effectively.

Table 2. Comparison of efficiency in production with and without AI

Efficiency Metric	With AI Integration	Without AI Integration
Crop Yield	Higher yield due to optimized growing conditions.	Lower yield with more variability in crop growth.
Water Usage	Reduced water usage through precise management.	Higher water consumption due to manual monitoring.
Energy Efficiency	Improved energy efficiency by optimizing lighting and temperature controls.	Higher energy usage due to less precise control.
Labor Requirements	Reduced need for labor, especially for monitoring and adjustments.	Higher labor costs for manual monitoring and adjustments.

continued on following page

Table 2. Continued

Efficiency Metric	With AI Integration	Without AI Integration
System Maintenance	Predictive maintenance reduces downtime and operational disruptions.	More frequent maintenance required, leading to potential downtimes.
Fish Health	Improved fish health through consistent water quality management.	Variability in fish health due to manual checks and adjustments.

FUTURE DIRECTIONS IN AI-ENHANCED AQUAPONICS

These emerging technologies and innovations can be seen to a great extent as surmises for the future of AI-enhanced aquaponics, likely to drive advances in sustainability and global food security. Thus, the chapter reviews some of the most interesting potential impacts stemming from emerging technologies such as the Internet of Things and blockchain, AI-driven innovation in sustainability, and the future of AI in food security (Agrawal, Magulur, et al., 2023; Agrawal, Shashibhushan, et al., 2023; P. R. Kumar et al., 2023; Pachiappan et al., 2024; Sankar et al., 2023; Venkateswaran et al., 2024).

Emerging Technologies: IoT, Blockchain, and AI

Internet of Things: The concept of the Internet of Things is sure to drastically transform AI-driven aquaponics with next-generation connectivity and real-time data collection capabilities. IoT devices, including sensors and smart controllers, track the quality of water, nutrient levels, environmental conditions, and fish health all the time. The devices communicate seamlessly with the data and use it in optimizing system performance by making adjustments automatically to the AI systems. IoT integration allows the increase of accuracy and speed in aquaponics operations, thus allowing more responsive and adaptive management.

Blockchain technology holds immense potential to enhance the transparency, traceability, and security of aquaponics systems. The technology ensures that information regarding system performance, resources being used, and all production processes are secure and verifiable through a decentralized and immutable ledger keeping track of data transactions. In this regard, it can increase the transparency of supply chains and facilitate the traceability of produce from the farm to the table, thereby gaining the trust of consumers and other stakeholders. In addition, blockchain smart contracts can automate and enforce agreements regarding the use of resources and supply chain transactions.

Advanced AI and Machine Learning: Advances in AI and machine learning for innovation are never stopped in an aquaponics system. New AI algorithms, notably reinforcement learning and deep learning, have advanced capabilities to optimize system operations and forecast future trends. Reinforcement learning algorithms learn from continuous interaction with the environment to improve the efficiency of the system over time. Deep learning models are capable of distilling out the complex patterns in data to provide insights pertinent to plant and fish health, resource use, and system performance, thus defining more precise and effective management strategies.

Innovations in AI-Driven Sustainability

Energy-Efficient Systems: Future innovations in AI will reside in continued minimization of energy use by aquaponics systems. AI-driven energy management systems will create an overview of heating, cooling, and lighting to ensure that the minimum energy is being used while retaining optimal conditions for plant and fish growth. Advances in energy storage technologies paired with AI will provide an avenue for managing renewable energy sources and hence improve aquaponics operations.

Water Recycling and Conversation: AI will help develop new water recycling and conservation technologies. With AI-driven innovations in filtering and water management systems, there will be increased efficiency in the recycling of water resources for sustainable use. AI will further enable real-time monitoring and manipulation of the parameters of water to prevent waste and enhance the efficiency of the entire system.

One of the major benefits of better resource management is that in years to come, artificial intelligence will still bring improvements in the use of nutrients, substrates, and other inputs. More sophisticated models of AI would be able to predict nutrient requirements much more effectively, thereby drastically reducing the use of synthetic fertilizers and finally easing the pressure on the environment. A new generation of AI-driven waste management systems will greatly raise the level of recycling of organic waste for the conversion of waste products into a valuable resource in the form of compost or biogas.

The Future of AI in Global Food Security

Large-Scale Food Production: AI-aided aquaponics can have a huge impact on food security globally by yielding sustainable and large-scale food production. AI-driven systems can help in developing large-scale aquaponics farms that produce high yields of fish and vegetables in an efficient manner and with little use of resources. All these farms can be found in various geographic regions—urban and arid regions—to ensure improved food security and decreased dependency on traditional agriculture.

Localized Food Systems: AI will play a major role in the growth of localized food systems by enabling the establishment of small-scale, urban aquaponics systems. They can grow fresh, healthy produce right next to cities, therefore cutting down on food transportation costs while decreasing carbon emissions related to food distribution. On this note, AI-driven optimization of the localized food systems will make them more efficient and sustainable, hence able to provide food security within large population centers.

Resilience to Climate Change: AI-enhanced aquaponics systems will play a critical role in building resilience to climate change. Through resource optimization and learning from environmental changes, AI will help the aquaponics systems endure shocks of climate variability such as extreme events and shifting patterns of temperature. This fast adaptation will sustain food production and thus food security globally amid a changing climate.

Global Collaboration and Knowledge Sharing: The future of AI in global food security will be characterized by increased collaboration and knowledge sharing among researchers, industry players, and policy makers. This type of collaboration will drive innovation, share best practices, and develop scalable solutions to food security challenges. AI platforms and networks will enable global cooperation and help spread the adoption of successful aquaponic practices and technologies across different regions.

Long-term AI-driven aquaponics will drive innovations in sustainability, efficiency, and global food security. Newer technologies such as IoT and blockchain will also link with other innovations about AI-driven sustainability, raising the performance and resilience of aquaponic systems. In view of challenges to the management of resources, energy use, and climate change, AI will be offered for scaleable and local food production with solutions to global food security. As AI technology is under further developments, its implementation in aquaponics will be very critical to the future of sustainable agriculture in light of emerging challenges affecting the world.

CONCLUSION

AI-enhanced aquaponics is one of the most radical innovations taking place in sustainable agriculture today, one that combines the principles of working aquaculture and hydroponics with state-of-the-art technology to enable global food security and environmental sustainability. In this chapter, we have gone through AI technologies integrated into aquaponics, showing their potentials in optimizing system performance, improving resource management, and facilitating innovative changes toward sustainability. AI technologies—machine learning, computer vision, and IoT—amount to a sea change in aquaponics monitoring and management. They provide for extremely fine-grained control over the environment; data analysis in real time; and automated adjustment of the system, thus increasing efficiency and productivity. Innovations such as AI-driven sustainability in energy use, water recycling, and advanced nutrient management demonstrate its role in reducing resource consumption while minimizing the impact on the environment.

Notwithstanding, some challenges and issues remain to be addressed by this field in terms of technological and financial issues, management of big data, and finally how human skills and AI integrate effectively. How the variety of challenges becomes addressed will lead either to a mass-scale adoption of AI-enhanced aquaponics systems or very limited success in this field. In the future, AI will play a significant role in scaling up aquaponics operations, localized food systems, and enhancing resilience towards climate change. Using emerging technologies and collaboration, AI can work towards a more sustainable, efficient, and secure food system globally.

At the Front of Agricultural Innovation, Promising Solutions to Some Very Serious Challenges Existing in Food Production and Environmental Sustainability. Its further evolution and adoption would be the key toward the fostering of sustainable agriculture and rising to the global food security challenge over the coming decades.

REFERENCES

Abbasi, R., Martinez, P., & Ahmad, R. (2022). An ontology model to represent aquaponics 4.0 system's knowledge. *Information Processing in Agriculture*, *9*(4), 514–532. DOI: 10.1016/j.inpa.2021.12.001

Agrawal, A. V., Magulur, L. P., Priya, S. G., Kaur, A., Singh, G., & Boopathi, S. (2023). Smart Precision Agriculture Using IoT and WSN. In *Handbook of Research on Data Science and Cybersecurity Innovations in Industry 4.0 Technologies* (pp. 524–541). IGI Global. DOI: 10.4018/978-1-6684-8145-5.ch026

Agrawal, A. V., Shashibhushan, G., Pradeep, S., Padhi, S., Sugumar, D., & Boopathi, S. (2023). Synergizing Artificial Intelligence, 5G, and Cloud Computing for Efficient Energy Conversion Using Agricultural Waste. In *Sustainable Science and Intelligent Technologies for Societal Development* (pp. 475–497). IGI Global.

Bali, M. K., & Singh, M. (2024). Farming in the Digital Age: AI-Infused Digital Twins for Agriculture. *2024 3rd International Conference on Sentiment Analysis and Deep Learning (ICSADL)*, 14–21.

Boopathi, S. (2024). Sustainable Development Using IoT and AI Techniques for Water Utilization in Agriculture. In *Sustainable Development in AI, Blockchain, and E-Governance Applications* (pp. 204–228). IGI Global. DOI: 10.4018/979-8-3693-1722-8.ch012

Boopathi, S., Kumar, P. K. S., Meena, R. S., Sudhakar, M., & Associates. (2023). Sustainable Dcvelopments of Modern Soil-Less Agro-Cultivation Systems: Aquaponic Culture. In *Human Agro-Energy Optimization for Business and Industry* (pp. 69–87). IGI Global.

Chandramenon, P., Aggoun, A., & Tchuenbou-Magaia, F. (2024). Smart approaches to Aquaponics 4.0 with focus on water quality- Comprehensive review. *Computers and Electronics in Agriculture*, *225*, 109256. DOI: 10.1016/j.compag.2024.109256

Channa, A. A., Munir, K., Hansen, M., & Tariq, M. F. (2024). Optimisation of Small-Scale Aquaponics Systems Using Artificial Intelligence and the IoT: Current Status, Challenges, and Opportunities. *Encyclopedia*, *4*(1), 313–336. DOI: 10.3390/encyclopedia4010023

Choi, E.-Y., Lee, H.-S., Cha, J. H., & Lee, L.-G. (2023). A Study on the AI Analysis of Crop Area Data in Aquaponics. *The Journal of the Convergence on Culture Technology*, *9*(3), 861–866.

Glady, J. B. P., D'Souza, S. M., Priya, A. P., Amuthachenthiru, K., Vikram, G., & Boopathi, S. (2024). A Study on AI-ML-Driven Optimizing Energy Distribution and Sustainable Agriculture for Environmental Conservation. In *Harnessing High-Performance Computing and AI for Environmental Sustainability* (pp. 1–27). IGI Global., DOI: 10.4018/979-8-3693-1794-5.ch001

Gnanaprakasam, C., Vankara, J., Sastry, A. S., Prajval, V., Gireesh, N., & Boopathi, S. (2023). Long-Range and Low-Power Automated Soil Irrigation System Using Internet of Things: An Experimental Study. In *Contemporary Developments in Agricultural Cyber-Physical Systems* (pp. 87–104). IGI Global.

Hui, L., Xiaochan, W., Yinyan, S., Tianpei, L., Haiming, Y., & Xiangze, H. (2022). Research status on aquaponics technology system. *Zhongguo Nong-Jihua Xuebao*, *43*(12), 75.

Jiang, W., & Liu, J. (2024). NeuroAqua: Developing an Optimized Artificial Intelligence and Internet of Things-based Aquaponics System. *2024 IEEE 3rd International Conference on Computing and Machine Intelligence (ICMI)*, 1–5.

Karimanzira, D., & Rauschenbach, T. (2021). An intelligent management system for aquaponics. *Automatisierungstechnik*, *69*(4), 345–350. DOI: 10.1515/auto-2020-0036

Koshariya, A. K., Kalaiyarasi, D., Jovith, A. A., Sivakami, T., Hasan, D. S., & Boopathi, S. (2023). AI-Enabled IoT and WSN-Integrated Smart Agriculture System. In *Artificial Intelligence Tools and Technologies for Smart Farming and Agriculture Practices* (pp. 200–218). IGI Global. DOI: 10.4018/978-1-6684-8516-3.ch011

Koshariya, A. K., Khatoon, S., Marathe, A. M., Suba, G. M., Baral, D., & Boopathi, S. (2023). Agricultural Waste Management Systems Using Artificial Intelligence Techniques. In *AI-Enabled Social Robotics in Human Care Services* (pp. 236–258). IGI Global. DOI: 10.4018/978-1-6684-8171-4.ch009

Kumar, P., Sampath, B., Kumar, S., Babu, B. H., & Ahalya, N. (2023). Hydroponics, Aeroponics, and Aquaponics Technologies in Modern Agricultural Cultivation. In *IGI:Trends, Paradigms, and Advances in Mechatronics Engineering* (pp. 223–241). IGI Global.

Kumar, P. R., Meenakshi, S., Shalini, S., Devi, S. R., & Boopathi, S. (2023). Soil Quality Prediction in Context Learning Approaches Using Deep Learning and Blockchain for Smart Agriculture. In *Effective AI, Blockchain, and E-Governance Applications for Knowledge Discovery and Management* (pp. 1–26). IGI Global. DOI: 10.4018/978-1-6684-9151-5.ch001

Kumara, V., Mohanaprakash, T., Fairooz, S., Jamal, K., Babu, T., & Sampath, B. (2023). Experimental Study on a Reliable Smart Hydroponics System. In *Human Agro-Energy Optimization for Business and Industry* (pp. 27–45). IGI Global. DOI: 10.4018/978-1-6684-4118-3.ch002

Liu, J., & Jiang, W. (2024). Optimization of Aquaponics System Efficiency based on Artificial Intelligence Approach. *2024 IEEE 3rd International Conference on Computing and Machine Intelligence (ICMI)*, 1–6.

Metin, A., Kasif, A., & Catal, C. (2023). Temporal fusion transformer-based prediction in aquaponics. *The Journal of Supercomputing*, *79*(17), 19934–19958. DOI: 10.1007/s11227-023-05389-8

Pachiappan, K., Anitha, K., Pitchai, R., Sangeetha, S., Satyanarayana, T., & Boopathi, S. (2024). Intelligent Machines, IoT, and AI in Revolutionizing Agriculture for Water Processing. In *Handbook of Research on AI and ML for Intelligent Machines and Systems* (pp. 374–399). IGI Global.

Rahamathunnisa, U., Subhashini, P., Aancy, H. M., Meenakshi, S., & Boopathi, S. (2023). Solutions for Software Requirement Risks Using Artificial Intelligence Techniques. In *Handbook of Research on Data Science and Cybersecurity Innovations in Industry 4.0 Technologies* (pp. 45–64). IGI Global.

Rozie, F., Syarif, I., & Al Rasyid, M. U. H. (2020). Design and implementation of intelligent aquaponics monitoring system based on IoT. *2020 International Electronics Symposium (IES)*, 534–540. DOI: 10.1109/IES50839.2020.9231928

Samikannu, R., Koshariya, A. K., Poornima, E., Ramesh, S., Kumar, A., & Boopathi, S. (2022). Sustainable Development in Modern Aquaponics Cultivation Systems Using IoT Technologies. In *Human Agro-Energy Optimization for Business and Industry* (pp. 105–127). IGI Global.

Sankar, K. M., Booba, B., & Boopathi, S. (2023). Smart Agriculture Irrigation Monitoring System Using Internet of Things. In *Contemporary Developments in Agricultural Cyber-Physical Systems* (pp. 105–121). IGI Global. DOI: 10.4018/978-1-6684-7879-0.ch006

Srinivas, B., Maguluri, L. P., Naidu, K. V., Reddy, L. C. S., Deivakani, M., & Boopathi, S. (2023). Architecture and Framework for Interfacing Cloud-Enabled Robots. In *Handbook of Research on Data Science and Cybersecurity Innovations in Industry 4.0 Technologies* (pp. 542–560). IGI Global. DOI: 10.4018/978-1-6684-8145-5.ch027

Styawati, S. (2023). Mustard Plant Growth Monitoring System Using Artificial Intelligence in Aquaponics. *IC-ITECHS*, *4*(1), 37–44.

Taha, M. F., ElMasry, G., Gouda, M., Zhou, L., Liang, N., Abdalla, A., Rousseau, D., & Qiu, Z. (2022). Recent advances of smart systems and internet of things (iot) for aquaponics automation: A comprehensive overview. *Chemosensors (Basel, Switzerland)*, *10*(8), 303. DOI: 10.3390/chemosensors10080303

Vangeri, A. K., Bathrinath, S., Anand, M. C. J., Shanmugathai, M., Meenatchi, N., & Boopathi, S. (2024). Green Supply Chain Management in Eco-Friendly Sustainable Manufacturing Industries. In *Environmental Applications of Carbon-Based Materials* (pp. 253–287). IGI Global., DOI: 10.4018/979-8-3693-3625-0.ch010

Venkateswaran, N., Kiran Kumar, K., Maheswari, K., Kumar Reddy, R. V., & Boopathi, S. (2024). Optimizing IoT Data Aggregation: Hybrid Firefly-Artificial Bee Colony Algorithm for Enhanced Efficiency in Agriculture. *AGRIS On-Line Papers in Economics and Informatics*, *16*(1), 117–130. DOI: 10.7160/aol.2024.160110

Vijaya Lakshmi, V., Mishra, M., Kushwah, J. S., Shajahan, U. S., Mohanasundari, M., & Boopathi, S. (2024). Circular Economy Digital Practices for Ethical Dimensions and Policies for Digital Waste Management. In *Harnessing High-Performance Computing and AI for Environmental Sustainability* (pp. 166–193). IGI Global., DOI: 10.4018/979-8-3693-1794-5.ch008

Zekrifa, D. M. S., Kulkarni, M., Bhagyalakshmi, A., Devireddy, N., Gupta, S., & Boopathi, S. (2023). Integrating Machine Learning and AI for Improved Hydrological Modeling and Water Resource Management. In *Artificial Intelligence Applications in Water Treatment and Water Resource Management* (pp. 46–70). IGI Global. DOI: 10.4018/978-1-6684-6791-6.ch003

KEY TERMS AND DEFINITIONS

AI: Artificial Intelligence.

CCPA: California Consumer Privacy Act.

CO: Carbon Monoxide.

DSS: Decision Support System.

GDPR: General Data Protection Regulation.

ML: Machine Learning.

Chapter 10
Commercial Applications of Aeroponics:
Revolutionizing Modern Agriculture and Sustainable Food Production

R. Vettriselvan
https://orcid.org/0000-0002-1324-136X
Academy of Maritime Education and Training, India

ABSTRACT

Aeroponics, a soilless cultivation technique, is gaining momentum as a transformative solution in modern agriculture. This chapter explores its commercial applications, focusing on its ability to revolutionize traditional farming and support sustainable food production. By delivering nutrients directly to plant roots in a controlled environment, aeroponics offers benefits like faster growth rates, higher yields, and reduced resource use. The chapter examines its adoption in urban farming, vertical farming, and greenhouse operations, along with its economic viability and environmental impact. Through case studies, the potential of aeroponics to shape the future of agriculture is highlighted, offering a practical overview of its commercial prospects.

DOI: 10.4018/979-8-3693-2320-5.ch010

1. INTRODUCTION

Agriculture, as the backbone of human civilization, has undergone significant transformations over the millennia. From the advent of traditional farming practices that harnessed natural ecosystems to the industrial agricultural revolution that prioritized efficiency and scale, the sector has continually evolved to meet the growing demands of a burgeoning global population. However, as we move further into the 21st century, agriculture faces unprecedented challenges. These include limited arable land, water scarcity, climate change, and the need for sustainable food production systems that can ensure food security for future generations. It is within this context that innovative farming techniques such as aeroponics have emerged as promising solutions to some of the most pressing issues in modern agriculture.

1.1. Overview of Aeroponics

Aeroponics is a soilless cultivation technique where plants are grown by suspending their roots in the air and periodically misting them with a nutrient-rich solution. Unlike traditional hydroponics, where plant roots are submerged in a water-based nutrient solution, aeroponics allows for more efficient oxygenation of the roots, leading to faster growth rates and higher yields. The concept of aeroponics was first introduced in the 1940s by W. Carter, who demonstrated the feasibility of growing plants in an air or mist environment without soil. However, it wasn't until the late 20th century that advancements in technology allowed for the practical application of aeroponics in commercial agriculture (Carter, 1942). The primary advantage of aeroponics lies in its resource efficiency. Since plants are grown without soil, the method significantly reduces the need for arable land, making it ideal for urban farming and regions with limited fertile land. Additionally, aeroponics uses up to 98% less water than traditional soil-based agriculture, making it a highly sustainable option in areas facing water scarcity (Despommier, 2011). Moreover, the closed-loop system used in aeroponics minimizes the need for chemical fertilizers and pesticides, reducing the environmental impact of agricultural production.

1.2. Significance in Modern Agriculture

The significance of aeroponics in modern agriculture cannot be overstated. As the global population continues to rise, with projections reaching nearly 10 billion by 2050, the demand for food is expected to increase by 70% (FAO, 2009). This surge in demand necessitates the development of innovative agricultural practices that can produce more food using fewer resources. Aeroponics, with its high efficiency and adaptability, offers a viable solution to this challenge. One of the key areas where aeroponics is making a significant impact is in urban agriculture. With more than half of the world's population now living in urban areas, there is a growing need to bring food production closer to consumers. Aeroponics allows for the cultivation of crops in urban settings, such as rooftops, vertical farms, and even inside buildings, thereby reducing the carbon footprint associated with transporting food from rural areas to cities (Benke & Tomkins, 2017). Furthermore, the use of aeroponics in urban farming can help alleviate food deserts—areas where access to fresh, healthy food is limited—by providing a local source of fresh produce. In addition to its role in urban agriculture, aeroponics is also gaining traction in regions with challenging environmental conditions. For instance, in arid regions where water is scarce, the water-efficient nature of aeroponics makes it an ideal solution for sustainable food production. Similarly, in areas with poor soil quality or contamination, aeroponics offers a way to grow crops without the need for soil, thereby bypassing issues related to soil degradation and pollution (Gopinath et al., 2020).

1.3. Technological Innovations and Advancements

The successful commercialization of aeroponics has been driven by significant technological advancements in recent years. One of the most critical innovations is the development of automated nutrient delivery systems, which ensure that plants receive the optimal amount of nutrients and water at precise intervals. These systems use sensors to monitor environmental conditions such as humidity, temperature, and light levels, allowing for real-time adjustments to the nutrient misting process (Kozai et al., 2016). This level of precision not only improves plant growth rates but also reduces resource wastage, further enhancing the sustainability of the aeroponic system. Another major technological breakthrough is the integration of aeroponics with the Internet of Things (IoT) and artificial intelligence (AI). IoT devices, such as smart sensors and actuators, can be used to collect data on various environmental parameters, which can then be analyzed by AI algorithms to optimize growing conditions. For example, AI can predict the best times to mist the plants based on weather forecasts or detect early signs of plant stress and adjust nutrient levels accordingly (Schiavon et al., 2020). These advancements not only increase the efficiency of aeroponic systems but also reduce the need for manual intervention, making them more accessible to a wider range of users, from small-scale urban farmers to large commercial operations. In addition to automation and AI, advancements in LED lighting technology have also played a crucial role in the growth of aeroponics. Traditional farming relies on natural sunlight, which can be inconsistent and limited by geographical location and season. However, modern aeroponic systems use energy-efficient LED lights that can be tuned to specific wavelengths to maximize photosynthesis and plant growth. This allows for year-round cultivation and the production of crops in regions that would otherwise be unsuitable for agriculture (Darko et al., 2014).

1.4. Economic and Environmental Impact

The economic and environmental impact of aeroponics is a subject of growing interest among researchers, policymakers, and industry stakeholders. On the economic front, aeroponics offers several advantages over traditional farming methods. The increased yield and faster growth rates associated with aeroponics can lead to higher profits for farmers, while the reduced need for land, water, and chemical inputs can lower production costs. Additionally, the ability to grow crops in urban areas can open up new markets and reduce the costs associated with transporting food from rural areas to cities (Despommier, 2011). From an environmental perspective, aeroponics has the potential to significantly reduce the ecological footprint of agriculture. The efficient use of water and nutrients in aeroponic systems can help conserve natural resources, while the elimination of soil erosion and runoff can reduce the impact of agriculture on ecosystems. Furthermore, the closed-loop nature of aeroponic systems minimizes the need for chemical fertilizers and pesticides, reducing the risk of pollution and promoting more sustainable farming practices (Gopinath et al., 2020). However, despite its many advantages, aeroponics is not without its challenges. The high initial cost of setting up an aeroponic system can be a barrier to entry for small-scale farmers, and the need for specialized knowledge and technical skills can limit its adoption. Moreover, the energy requirements of aeroponic systems, particularly those that rely on artificial lighting, can be significant, raising concerns about their long-term sustainability (Kozai et al., 2016).

1.5. Purpose and Scope

The purpose of this chapter is to provide an in-depth exploration of the commercial applications of aeroponics and its potential to revolutionize modern agriculture and sustainable food production. The chapter will examine the historical development of aeroponics, the technological innovations that have enabled its commercialization, and its economic and environmental impact. Additionally, the chapter will explore the challenges and limitations of aeroponics, as well as the future trends and research directions that could shape its development in the coming years. In doing so, this chapter aims to provide advanced readers with a comprehensive understanding of the current state of aeroponics and its potential to transform agriculture. The chapter will also offer practical insights and recommendations for policymakers, industry stakeholders, and researchers interested in advancing the adoption and development of aeroponic systems. The readers will have a thorough understanding of the commercial applications of aeroponics, its advantages and limitations, and the role it can play in addressing the global challenges of food security and sustainability. The chapter will also highlight the need for continued research and innovation in this field, particularly in areas such as energy efficiency, cost reduction, and the integration of emerging technologies.

2. COMPREHENSIVE LITERATURE REVIEW

2.1. Historical Evolution of Aeroponics

Aeroponics, a method of growing plants without the use of soil or aggregate medium, dates back to the early 20th century. Its conceptual origins can be traced to the work of botanists and agricultural scientists who sought to understand plant root systems in more controlled environments. The foundational principles of aeroponics were first introduced by researchers such as W. Carter in 1942, who developed a method to grow plants in an air or mist environment without soil, facilitating the study of plant roots in a more natural state (Carter, 1942). This early work laid the groundwork for future developments in aeroponics, marking the beginning of a revolutionary approach to plant cultivation. The term "aeroponics" itself was coined later, during the 1960s, as the practice began to gain traction among researchers and experimental horticulturists. The 1980s saw significant advancements in the field, particularly with the development of high-pressure misting systems, which allowed for more precise and consistent delivery of nutrient-rich mist to the plant roots (Goto, 2012). This period also witnessed the first commercial applications of aeroponics, particularly in the cultivation of high-value crops such as strawberries and lettuce, where the benefits of controlled environment agriculture could be fully realized (Soffer & Burger, 1988). The evolution of aeroponics has been marked by continuous innovation and refinement, driven by the increasing need for sustainable and efficient agricultural practices. The 1990s and 2000s saw the integration of computer-controlled systems, which allowed for more precise regulation of environmental conditions such as humidity, temperature, and nutrient concentration (Stoner & Lewis, 2003). These technological advancements significantly improved the reliability and scalability of aeroponic systems, making them more accessible to commercial growers and researchers alike. In recent years, the focus of aeroponics has expanded beyond traditional agriculture to include applications in space exploration and urban farming. NASA has conducted extensive research into the use of aeroponics for growing food in space, as part of its efforts to develop sustainable life support systems for long-duration space missions (Mitchell, 1996). The ability to grow fresh produce in space, without the need for soil or large amounts of water, makes aeroponics an ideal solution for space habitats, where resources are limited, and environmental control is paramount. Urban farming, another emerging application of aeroponics, leverages the system's ability to produce high yields in a small footprint, making it particularly suited for densely populated urban areas. Vertical farms, which utilize aeroponic systems to grow crops in stacked layers, are becoming increasingly popular as a means of producing fresh, local food in cities where traditional agriculture is not feasible (Despommier, 2011). This trend reflects a broader shift towards more sustainable and localized food production systems, driven by growing concerns over food security, climate change, and environmental sustainability.

2.2. Technological Innovations in Aeroponics

The success and scalability of aeroponics have been largely driven by technological innovations that have improved the efficiency and reliability of the system. One of the most significant advancements in aeroponics is the development of sophisticated nutrient delivery systems, which provide plants with a precise mixture of water, nutrients, and oxygen directly to their roots. Early aeroponic systems used simple misting techniques, but modern systems have evolved to include high-pressure pumps and nozzles that produce fine aerosols, ensuring that the plant roots are evenly coated with nutrients (Goto, 2012). These advancements in nutrient delivery have been complemented by the integration of automation and environmental controls, which allow for the precise regulation of the growing environment. Automated systems can monitor and adjust variables such as temperature, humidity, light intensity, and nutrient levels in real-time, optimizing the conditions for plant growth. This level of control is particularly important in aeroponics, where small fluctuations in environmental conditions can have a significant impact on plant health and productivity (Stoner & Lewis, 2003). The use of sensors and computer-controlled systems has made it possible to maintain optimal growing conditions consistently, reducing the risk of crop failure and improving overall yields. Another key innovation in aeroponics is the use of LED lighting, which has become the preferred light source for indoor and vertical farms. LED lights are energy-efficient, produce little heat, and can be tailored to emit specific wavelengths of light that are most beneficial for photosynthesis. This allows growers to optimize the light spectrum for different stages of plant growth, further enhancing the efficiency of the system (Darko et al., 2014). Additionally, the ability to control the light cycle precisely enables year-round production, irrespective of the season or weather conditions. In recent years, the integration of big data analytics and artificial intelligence (AI) into aeroponic systems has opened new possibilities for precision agriculture. By collecting and analyzing large datasets on plant growth, nutrient uptake, and environmental conditions, AI algorithms can identify patterns and optimize growing conditions in real-time (Schiavon et al., 2020). This not only improves the efficiency of aeroponic systems but also enables predictive modeling, which can help growers anticipate and mitigate potential issues before they impact crop yields. The ongoing development of modular and scalable aeroponic systems is also worth noting. These systems are designed to be easily expandable, allowing growers to increase production capacity as needed without significant additional investment. This flexibility makes aeroponics an attractive option for both small-scale urban farms and large commercial operations, as it can be tailored to meet the specific needs and constraints of the grower (Benke & Tomkins, 2017).

2.3. Economic and Environmental Impact

The economic and environmental impact of aeroponics has been the subject of numerous studies, many of which highlight the system's potential to address some of the most pressing challenges facing modern agriculture. One of the most significant economic advantages of aeroponics is its ability to produce high yields in a relatively small footprint. Because plants are grown vertically and without soil, aeroponic systems can be installed in areas where traditional agriculture is not feasible, such as urban environments or arid regions (Despommier, 2011). This reduces the need for arable land and makes it possible to produce food closer to where it is consumed, reducing transportation costs and associated carbon emissions. Aeroponics is also highly resource-efficient, particularly in terms of water usage. Traditional agriculture is notoriously water-intensive, with irrigation accounting for a significant proportion of global freshwater consumption. In contrast, aeroponics uses up to 95% less water than conventional farming methods, as the nutrient solution is delivered directly to the plant roots and recirculated within the system (Schiavon et al., 2020). This makes aeroponics an attractive option for regions facing water scarcity, where the ability to conserve water while maintaining high agricultural productivity is critical.

In addition to its water-saving benefits, aeroponics also reduces the need for chemical inputs such as pesticides and herbicides. Because the plants are grown in a controlled environment, they are less susceptible to pests and diseases, reducing the need for chemical treatments (Gopinath et al., 2020). This not only lowers the cost of production but also minimizes the environmental impact of farming, as the use of synthetic chemicals in agriculture is a major source of pollution and biodiversity loss. The potential for year-round production is another economic benefit of aeroponics. Unlike traditional farming, which is often limited by seasonal variations, aeroponics can produce crops continuously throughout the year, providing a steady supply of fresh produce. This is particularly important for high-value crops such as leafy greens, herbs, and berries, which can command premium prices in the market (Benke & Tomkins, 2017). The ability to produce food year-round also enhances food security, as it reduces dependence on seasonal imports and mitigates the impact of climate-related disruptions to agricultural production. However, it is important to note that the economic viability of aeroponics is not without its challenges. The initial capital investment required to set up an aeroponic system can be significant, particularly for large-scale operations. The cost of purchasing and installing the necessary equipment, such as misting systems, nutrient reservoirs, and environmental controls, can be prohibitive for small-scale farmers or those in developing regions (Benke & Tomkins, 2017). Additionally, the ongoing maintenance and energy costs associated with operating an aeroponic system can further add to the financial burden. Despite these challenges, the long-term economic benefits of aeroponics, particularly in terms of resource efficiency and year-round production, make it an attractive option for sustainable agriculture. As technology continues to advance and the cost of equipment decreases, it is likely that the adoption of aeroponics will continue to grow, particularly in regions where traditional agriculture is not feasible or sustainable.

2.4. Comparative Analysis of Other Soilless Systems

Aeroponics is often compared with other soilless cultivation systems, such as hydroponics and aquaponics, as well as traditional soil-based agriculture. Each of these systems has its own advantages and limitations, and the choice of system often depends on factors such as the type of crop being grown, the available resources, and the specific goals of the grower. Hydroponics, like aeroponics, involves growing plants without soil, but instead of suspending the roots in the air, hydroponic systems use a nutrient-rich water solution to support plant growth. Hydroponics is widely used in commercial agriculture and has been shown to produce high yields with less water and space than traditional farming methods (Resh, 2013). However, hydroponic systems can be more susceptible to diseases and require careful monitoring of water quality and nutrient levels to prevent crop failure. Aquaponics combines hydroponics with aquaculture, using fish waste to provide nutrients for the plants. This system creates a symbiotic relationship between the plants and the fish, with the plants helping to filter and clean the water for the fish, while the fish waste provides a natural source of nutrients for the plants. Aquaponics is praised for its sustainability, as it recycles water and nutrients within a closed-loop system, minimizing waste and the need for external inputs (Love et al., 2015). However, the complexity of managing both plant and fish health can be a drawback, especially for those without expertise in aquaculture. Additionally, aquaponics systems typically require more space and are less scalable than aeroponics, making them less suitable for urban environments or high-density farming.

When comparing aeroponics to traditional soil-based agriculture, the differences are even more pronounced. Soil-based farming relies heavily on natural soil fertility and climatic conditions, which can be highly variable and difficult to control. This variability often leads to lower yields and higher susceptibility to pests, diseases, and environmental stressors such as drought or flooding (Altieri, 2018). In contrast, aeroponics offers a highly controlled environment where variables such as temperature, humidity, and nutrient delivery can be precisely managed to optimize plant growth. This control not only leads to higher yields but also allows for the cultivation of crops in regions where traditional agriculture would be impossible, such as arid deserts or urban centers. Another important consideration is the environmental impact of each system. Traditional agriculture is a major contributor to environmental degradation, through processes such as soil erosion, water pollution from pesticide and fertilizer runoff, and deforestation for the expansion of farmland (Foley et al., 2011). In contrast, aeroponics and other soilless systems have a much lower environmental footprint. As previously mentioned, aeroponics uses significantly less water and does not require the use of soil, which reduces the need for land and minimizes soil degradation. Additionally, the controlled environment of aeroponics reduces the need for chemical inputs, leading to fewer pollutants entering the environment.

However, the sustainability of aeroponics must also be evaluated in terms of energy use. Aeroponic systems are highly dependent on electricity to power pumps, lights, and environmental controls. In regions where electricity is generated from fossil fuels, this energy use can offset some of the environmental benefits of aeroponics (Benke & Tomkins, 2017). Therefore, the adoption of renewable energy sources, such as solar or wind power, is essential to maximize the sustainability of aeroponic farming. While aeroponics offers several advantages over both traditional and other soilless farming methods—such as higher yields, resource efficiency, and the ability to grow crops in challenging environments—each system has its own set of trade-offs. The choice between aeroponics, hydroponics, aquaponics, or traditional agriculture ultimately depends on the specific needs and resources of the grower, as well as the environmental and economic context in which the system is being implemented.

3. CASE STUDIES AND REAL-WORLD APPLICATIONS

3.1. Successful Commercial Aeroponics Operations

In recent years, the commercial adoption of aeroponics has seen significant growth, with several companies demonstrating its economic viability and sustainability. One of the most notable examples is AeroFarms, a company based in Newark, New Jersey. Founded in 2004, AeroFarms has developed one of the world's largest aeroponic vertical farms, producing a variety of leafy greens. Their system utilizes a closed-loop irrigation system that recirculates water, reducing water usage by up to 95% compared to traditional agriculture (Banerjee & Adenaeuer, 2014). AeroFarms has also integrated LED lighting to optimize photosynthesis, allowing plants to grow faster and with higher nutrient content. The economic outcomes of AeroFarms have been equally impressive. The company reports yields up to 390 times higher per square foot per year compared to conventional field farming (Despommier, 2020). This efficiency is partly due to the ability to stack plants in vertical layers, maximizing space use in urban environments. Furthermore, AeroFarms has successfully tapped into the local food movement by providing fresh produce to nearby urban areas, reducing transportation costs and carbon emissions associated with food distribution (Kozai, Niu, & Takagaki, 2016).

Another successful example is LettUs Grow, a UK-based company specializing in aeroponic systems for both commercial and research purposes. Their proprietary technology, called aeroponic misting, uses an ultrasonic mist to deliver nutrients directly to plant roots, which is more efficient than traditional spray systems. LettUs Grow's systems are designed to be modular, making them scalable for different sizes of operations, from small research labs to large commercial farms (Manning & Chinnery, 2021). This flexibility has allowed LettUs Grow to cater to a diverse range of clients, including educational institutions, research centers, and commercial growers. The economic and environmental benefits of LettUs Grow's technology have been recognized through various awards and grants. The company has received funding from Innovate UK and the European Space Agency to further develop its systems for use in space and other extreme environments, highlighting the potential of aeroponics for future agricultural challenges (Harbaugh, 2022). These case studies illustrate the commercial potential of aeroponics when combined with innovative technology and strategic market positioning.

3.2. Urban and Vertical Farming Initiatives

Aeroponics has also been increasingly adopted in urban and vertical farming initiatives, where space is limited, and traditional farming methods are not feasible. Sky Greens, a Singapore-based vertical farm, is a pioneering example of how aeroponics can be applied in densely populated urban environments. Sky Greens employs an innovative vertical farming system that uses low-carbon hydraulic water-driven systems to rotate plant racks, ensuring even light distribution and maximizing space efficiency. The farm produces a variety of vegetables that are sold directly to local supermarkets, reducing the carbon footprint associated with food transportation (Bayati, 2018). Sky Greens' approach to vertical farming is both sustainable and economically viable. The farm uses 95% less water, 75% less fertilizer, and up to 90% less energy compared to traditional farming methods. This is achieved through the combination of aeroponics and a closed-loop water system, which recycles water and nutrients. The integration of renewable energy sources, such as solar panels, further reduces the farm's reliance on non-renewable energy, making it a model for sustainable urban agriculture (Lim et al., 2020).

Another noteworthy urban farming initiative is Plenty, a California-based company that specializes in vertical aeroponic farming. Plenty's farms are designed to operate in urban areas, close to where food is consumed. The company's technology uses a combination of sensors, machine learning, and robotics to monitor and optimize growing conditions in real time, ensuring optimal plant growth and resource efficiency. This high-tech approach allows Plenty to produce large quantities of fresh produce year-round, even in areas with harsh climates or limited arable land (Ming, 2021). Plenty's focus on urban farming not only addresses the issue of food security in cities but also contributes to reducing the environmental impact of agriculture. By growing food close to consumers, Plenty reduces the need for long-distance transportation, which is a significant source of greenhouse gas emissions. Additionally, the company's farms use 99% less land and 95% less water than traditional farming, making them a sustainable solution for urban food production (Despommier, 2020).

3.3. Integration with Emerging Technologies

The integration of emerging technologies with aeroponics is driving the next wave of innovation in sustainable agriculture. The Internet of Things (IoT) plays a crucial role in the automation and optimization of aeroponic systems. IoT devices can monitor environmental conditions such as temperature, humidity, and nutrient levels in real-time, allowing farmers to make data-driven decisions to optimize plant growth (Raviv & Lieth, 2019). For example, sensors placed within the aeroponic system can detect when plants need more nutrients or water, triggering automated systems to deliver the necessary resources without human intervention. Artificial Intelligence (AI) is another technology that is being increasingly integrated into aeroponics. AI algorithms can analyze vast amounts of data collected by IoT devices to predict plant growth patterns, identify potential issues before they become problems, and optimize resource use. Companies like Iron Ox are leading the way in this area, using AI-driven robots to manage their aeroponic farms. These robots can plant, tend, and harvest crops autonomously, significantly reducing labor costs and increasing efficiency (Goldstein, 2021).

Blockchain technology is also finding its place in aeroponics, particularly in ensuring transparency and traceability in the food supply chain. Blockchain can be used to track the journey of produce from farm to table, providing consumers with information about where their food comes from and how it was grown. This level of transparency is increasingly demanded by consumers who are concerned about food safety and environmental sustainability (Sundmaeker, Verdouw, Wolfert, & Pérez Freire, 2016). The integration of these technologies with aeroponics is not only enhancing productivity and sustainability but also making it possible to scale up operations and bring aeroponic farming to a wider audience. As technology continues to evolve, we can expect to see even more innovative applications of aeroponics in the future, further solidifying its role as a key component of sustainable agriculture.

4. CHALLENGES AND LIMITATIONS

4.1. Technical Challenges

The widespread adoption of aeroponics faces several technical challenges that must be addressed to fully realize its potential. One of the primary technical barriers is the complexity of aeroponic systems. Unlike traditional soil-based farming, aeroponics relies on a precisely controlled environment where plants are suspended in the air, and their roots are misted with a nutrient-rich solution. This complexity necessitates advanced systems for monitoring and regulating various factors such as nutrient concentration, pH levels, humidity, temperature, and light (Raviv & Lieth, 2019). Failure to maintain these parameters within optimal ranges can lead to poor plant growth or even crop failure, making the technology less accessible to small-scale or novice farmers. Maintenance requirements also present a significant challenge. Aeroponic systems are highly sensitive to operational disruptions, such as clogging of mist nozzles, pump failures, or power outages. These issues can lead to a sudden drop in productivity or complete crop loss if not promptly addressed. Regular maintenance is necessary to ensure the smooth functioning of the system, which includes cleaning and replacing components like nozzles, pumps, and filters. The need for such frequent and specialized maintenance can be a deterrent for potential adopters who may not have the technical expertise or resources to manage these systems effectively (Li et al., 2020). Energy consumption is another critical technical challenge associated with aeroponics. The systems rely heavily on electrical power to run pumps, lighting, and climate control mechanisms. While energy-efficient technologies such as LED lights and energy recovery systems have been developed to mitigate this issue, the overall energy demand of aeroponic farms remains high compared to traditional farming methods (Benke & Tomkins, 2017). In regions where electricity costs are high or where the energy grid is unreliable, this can significantly impact the economic viability of aeroponic farming. Moreover, the reliance on non-renewable energy sources for powering these systems can negate some of the environmental benefits of aeroponics, such as reduced water usage and pesticide-free farming.

4.2. Economic Barriers

Economic challenges represent a significant barrier to the broader adoption of aeroponics. One of the most prominent economic barriers is the high initial setup cost. Establishing an aeroponic farm involves significant capital investment in infrastructure, including the construction of controlled environment structures, acquisition of aeroponic equipment, and installation of energy systems (Kozai, Niu, & Takagaki, 2016). The cost of these components can be prohibitively expensive, particularly for small-scale farmers or startups. Additionally, there is a need for continuous investment in the form of maintenance, energy costs, and system upgrades, which can further strain the financial resources of farmers. Market acceptance of aeroponically grown produce also poses an economic challenge. While the demand for locally grown, pesticide-free food has been increasing, consumers and retailers may still have reservations about the quality and safety of aeroponically grown crops. This is partly due to the lack of familiarity with the technology and misconceptions about the nutritional value of produce grown without soil (Grewal, 2021). Overcoming this market resistance requires significant marketing efforts and consumer education, which can add to the overall cost burden for aeroponic farmers. Furthermore, the adoption of aeroponics requires specialized knowledge and skills. The operation of aeroponic systems is more complex than traditional farming, requiring an understanding of plant physiology, nutrient management, and system maintenance. This creates a barrier to entry for individuals or organizations without access to such expertise. The need for training and education in aeroponic techniques can be a significant economic hurdle, particularly in regions where agricultural education and technical support are limited (Zou et al., 2022). This lack of accessible knowledge can also lead to operational inefficiencies and increased costs, as farmers may struggle to optimize their systems or troubleshoot issues.

4.3. Regulatory and Policy Issues

Regulatory and policy challenges also play a crucial role in shaping the adoption and expansion of aeroponics. The regulatory landscape for aeroponics is still evolving, and there are significant variations in how different regions and countries govern this technology. One of the primary regulatory issues is the classification of aeroponic produce. In some regions, aeroponically grown crops may not be recognized under existing organic certification standards because they are not grown in soil, which is traditionally a requirement for organic labeling (Goldstein, 2021). This lack of certification can limit market opportunities for aeroponic farmers who wish to target the growing organic food market. Moreover, the regulatory approval process for new agricultural technologies, including aeroponics, can be lengthy and cumbersome. This is particularly challenging in regions with stringent agricultural regulations or where government agencies are slow to adapt to new farming methods. The approval process often involves rigorous testing and evaluation to ensure that the technology is safe for both consumers and the environment. Delays in obtaining regulatory approval can hinder the development and commercialization of aeroponic systems, making it difficult for innovators and entrepreneurs to bring new products to market (Ruan et al., 2021).

Policy challenges also extend to the availability of financial incentives and support for aeroponic farming. In many regions, agricultural subsidies and grants are still predominantly directed towards traditional farming practices, leaving aeroponic farmers without the financial support needed to offset the high costs of adopting this technology. The lack of targeted policies and incentives for sustainable farming practices like aeroponics can slow down the transition to more environmentally friendly and efficient food production systems (Sundmaeker, Verdouw, Wolfert, & Pérez Freire, 2016). Policymakers need to recognize the potential of aeroponics and other innovative farming techniques in addressing global food security challenges and climate change, and they must create a supportive regulatory and policy framework to encourage their adoption. While aeroponics offers numerous benefits, including reduced water usage, pesticide-free farming, and the potential for high yields in controlled environments, several challenges must be addressed to enable its widespread adoption. Technical challenges such as system complexity, maintenance requirements, and energy consumption, coupled with economic barriers like high initial costs and market acceptance, pose significant obstacles. Additionally, the regulatory and policy landscape needs to evolve to support the growth of aeroponics and other innovative agricultural practices. Addressing these challenges will require concerted efforts from researchers, policymakers, industry stakeholders, and farmers to create a more sustainable and resilient food production system.

5. FUTURE TRENDS AND RESEARCH DIRECTIONS

5.1. Innovations in System Design

Aeroponics is on the brink of significant technological evolution, driven by advancements in system design aimed at improving operational efficiency and reducing costs. One major innovation is the development of advanced materials for aeroponic systems. Research is focusing on the use of high-performance polymers and composites, which offer enhanced durability and resistance to environmental factors. For instance, studies have demonstrated that incorporating nanomaterials into system components can significantly improve their mechanical strength and lifespan (Nielsen et al., 2024). Additionally, these materials often exhibit better resistance to degradation from nutrient solutions and ultraviolet (UV) radiation, leading to longer-lasting and more reliable systems. Another cutting-edge advancement is the integration of sophisticated control systems using artificial intelligence (AI) and machine learning (ML). These systems can analyze vast amounts of data from sensors embedded in the aeroponic setup to optimize growing conditions in real time. For example, AI algorithms can predict nutrient deficiencies, adjust pH levels, and control environmental parameters such as light and temperature with high precision (Brown & Liu, 2024). This level of automation not only enhances plant growth and yields but also reduces labor costs and the risk of human error. Moreover, the advent of modular and scalable aeroponic systems represents a significant shift towards flexibility and adaptability. Modular designs allow for easy expansion and reconfiguration, catering to diverse applications from small-scale urban farms to large commercial operations. Research has shown that modular systems can be designed to fit various spatial constraints and operational scales, making them suitable for a wide range of environments (Gonzalez et al., 2024). These systems can also be integrated with other agricultural technologies, such as hydroponics and vertical farming, to create hybrid growing environments that maximize resource efficiency.

5.2. Sustainable Practices and Green Technologies

Sustainability is increasingly becoming a central focus in aeroponic farming, with research exploring various ways to integrate green technologies and practices. One notable trend is the incorporation of renewable energy sources to power aeroponic systems. Solar energy, for instance, is being harnessed through photovoltaic panels installed on the structures housing aeroponic systems. Studies have demonstrated that solar-powered aeroponic systems can significantly reduce reliance on grid electricity and lower operational costs (Jones & Smith, 2023). Similarly, wind energy is being explored as a complementary power source, particularly in areas with consistent wind patterns. Waste recycling and closed-loop systems are also pivotal in enhancing the sustainability of aeroponics. Researchers are developing advanced filtration and purification systems to recycle nutrient solutions and water. These systems capture runoff, filter out contaminants, and return the cleaned solution to the system, minimizing waste and reducing the need for fresh inputs (Miller & Thompson, 2024). Such practices not only conserve resources but also mitigate environmental impacts associated with traditional agricultural practices. The application of biodegradable and recyclable materials in aeroponic systems is another important aspect of sustainability. Research is focusing on the development of eco-friendly materials that can replace conventional plastics used in system components. For example, bioplastics derived from plant-based sources are being tested for use in aeroponic structures, offering a more sustainable alternative that reduces plastic waste (Nguyen et al., 2023). Additionally, organic and non-synthetic nutrient sources are being integrated into aeroponic systems to further enhance their environmental sustainability.

5.3. Global Adoption and Market Trends

The global adoption of aeroponics is expanding rapidly, driven by increasing awareness of its benefits and advancements in technology. Market research indicates that the aeroponics industry is expected to grow substantially, with projections estimating a market size of USD 5 billion by 2028, reflecting a compound annual growth rate (CAGR) of over

12% (MarketWatch, 2024). This growth is attributed to several factors, including the rising demand for efficient and sustainable food production methods, technological advancements, and supportive policy frameworks. Government and private sector initiatives are instrumental in fostering the adoption of aeroponics. Various governments are implementing policies and providing funding to support the development and deployment of aeroponic technologies. For instance, grant programs and subsidies are available to encourage research and investment in advanced agricultural technologies (European Commission, 2024). Additionally, private sector investments are driving innovation and commercialization, with numerous startups and established companies entering the aeroponics market and contributing to its expansion.

Educational and outreach programs play a critical role in accelerating global adoption. As more stakeholders, including farmers, investors, and policymakers, become informed about the benefits and potential of aeroponics, there is a growing interest in implementing these systems. Conferences, workshops, and educational initiatives are helping to disseminate knowledge, share best practices, and promote the adoption of aeroponics on a global scale (Smith et al., 2023). This increased awareness is expected to drive further investment and innovation in the field. The future of aeroponics is marked by significant advancements in system design, sustainability, and global adoption. Innovations in materials, smart technologies, and modular systems are set to enhance the efficiency and adaptability of aeroponic systems. The integration of renewable energy, waste recycling, and sustainable practices will contribute to the environmental sustainability of these systems. The global market for aeroponics is poised for substantial growth, supported by government and private sector initiatives and increasing awareness among stakeholders. These trends and research directions will shape the future of aeroponics, revolutionizing modern agriculture and contributing to sustainable food production.

6. RECOMMENDATIONS FOR ADVANCING AEROPONICS

6.1. Strategies for Scaling Aeroponics

Scaling aeroponics from small-scale operations to large commercial farms involves addressing several critical factors, including system design, resource management, and market access. Here are key strategies to facilitate this transition:

i. **Modular and Scalable Designs:** Modular aeroponic systems are essential for scaling operations effectively. These systems allow for incremental expansion by adding new modules or units as needed. Research suggests that modularity enhances flexibility, reduces initial capital expenditure, and simplifies system upgrades (Gonzalez et al., 2024). By adopting modular designs, operators can start with a smaller setup and gradually scale up as demand increases.

ii. **Standardization and Automation:** Standardizing system components and integrating automation are crucial for scaling aeroponic farms. Standardized components streamline manufacturing, reduce costs, and ensure compatibility across different system modules. Automation technologies, including robotic systems and AI-driven controls, improve efficiency and consistency in larger operations (Smith & Liu, 2023). Automated systems can manage nutrient delivery, environmental controls, and monitoring with minimal human intervention, which is vital for large-scale farms.

iii. **Economies of Scale:** Achieving economies of scale is fundamental to reducing costs and enhancing profitability in large aeroponic operations. Bulk purchasing of materials, centralized management of resources, and optimized logistics can significantly lower operational costs. Studies indicate that larger-scale aeroponic systems benefit from reduced per-unit costs for inputs such as nutrients, energy, and labor (Brown & Patel, 2024). Implementing these practices can make large-scale aeroponics more competitive compared to traditional farming methods.

iv. **Integration with Existing Agricultural Practices:** Integrating aeroponics with existing agricultural practices can enhance scalability. For instance, combining aeroponics with vertical farming or hydroponics can create hybrid systems that optimize resource use and increase production capacity. Such integrations can also facilitate the transition from small-scale to large-scale operations by leveraging existing infrastructure and expertise (Jones & Smith, 2023).

v. **Investment in Research and Development:** Investing in research and development (R&D) is critical for advancing aeroponics technology and scaling operations. R&D efforts should focus on improving system efficiency, reducing costs, and developing innovative solutions for large-scale implementation. Collaboration between industry stakeholders, academic institutions, and government agencies can drive technological advancements and support the growth of aeroponic systems (Lee & Kim, 2023).

6.2. Enhancing Economic Viability

Improving the economic viability of aeroponics involves implementing cost-reduction techniques, securing government support, and developing market strategies. Here are strategies to enhance economic viability:

i. **Cost-Reduction Techniques:** Reducing operational costs is essential for improving the economic viability of aeroponics. Techniques such as optimizing nutrient use, reducing energy consumption, and improving system efficiency can lower costs. For instance, advancements in LED lighting technology offer energy-efficient solutions for aeroponic systems, reducing electricity costs while enhancing plant growth (Garcia et al., 2023). Additionally, implementing closed-loop nutrient recycling systems can minimize waste and reduce the need for external nutrient inputs (Miller & Thompson, 2024).

ii. **Government Subsidies and Incentives:** Government subsidies and incentives can significantly impact the economic viability of aeroponics. Financial support for research, development, and deployment of aeroponic systems can reduce initial capital costs and encourage investment. Governments can offer grants, tax credits, and low-interest loans to support the adoption of aeroponics in commercial agriculture (European Commission, 2024). Such incentives can make aeroponics more financially attractive and accelerate its widespread adoption.

iii. **Market Development and Expansion:** Developing and expanding markets for aeroponic products is crucial for enhancing economic viability. Establishing partnerships with retailers, food distributors, and restaurants can create new market opportunities for aeroponically grown produce. Additionally, educating consumers about the benefits of aeroponically grown food, such as higher nutritional value and reduced environmental impact, can drive demand and increase market share (Smith et al., 2023). Expanding into niche markets, such as organic and specialty produce, can also enhance profitability.

iv. **Technology Transfer and Knowledge Sharing:** Facilitating technology transfer and knowledge sharing among stakeholders can improve economic viability. By sharing best practices, technical expertise, and innovative solutions, stakeholders can enhance the efficiency and effectiveness of aeroponic systems. Industry conferences, workshops, and online platforms can serve as forums for exchanging knowledge and fostering collaboration (Brown & Liu, 2024). Such initiatives can help reduce costs and improve the overall economic performance of aeroponic operations.

6.3. Policy and Regulatory Support

Supportive policy measures and regulatory frameworks are essential for promoting the growth of aeroponics at both national and international levels. Here are key recommendations for policy and regulatory support:

i. **Developing Supportive Policies:** Governments should develop and implement policies that support the growth of aeroponics. These policies could include financial incentives, research funding, and regulatory frameworks that facilitate the adoption of aeroponic technologies. Policies that encourage sustainable agricultural practices and support innovation in food production can create a favorable environment for aeroponics (European Commission, 2024). Additionally, policies that streamline the approval process for new technologies and products can accelerate the deployment of aeroponic systems.

ii. **Creating Standards and Regulations:** Establishing standards and regulations for aeroponics is crucial for ensuring system quality, safety, and consistency. Regulatory frameworks should address key aspects such as system design, nutrient management, and environmental controls. Clear guidelines for the certification of aeroponic products and systems can help build consumer trust and facilitate market entry (Nguyen et al., 2023). Collaboration between industry stakeholders, regulatory agencies, and standardization bodies is essential for developing and implementing effective standards.

iii. **Encouraging International Collaboration:** International collaboration can enhance the growth and development of aeroponics by facilitating knowledge exchange, technology transfer, and market access. Governments and industry organizations should promote partnerships and collaborations between countries to share best practices, research findings, and technological innovations. International agreements and initiatives that support sustainable agriculture and advanced technologies can drive global adoption of aeroponics and contribute to its growth (Gonzalez et al., 2024).

iv. **Supporting Education and Training:** Investing in education and training programs is vital for developing a skilled workforce and promoting the adoption of aeroponics. Governments and educational institutions should offer training programs, certification courses, and workshops to equip individuals with the necessary skills and knowledge to operate and manage aeroponic systems. Educational initiatives should focus on both technical aspects of aeroponics and broader topics such as sustainability and resource management (Smith & Liu, 2023). A well-trained workforce can drive innovation and ensure the successful implementation of aeroponic technologies.

v. **Promoting Public-Private Partnerships:** Public-private partnerships (PPPs) can play a key role in advancing aeroponics by combining the strengths of both sectors. Governments can collaborate with private companies, research institutions, and non-governmental organizations to support research, development, and commercialization of aeroponic technologies. PPPs can facilitate the sharing of resources, expertise, and funding, leading to accelerated innovation and market growth (Jones & Smith, 2023). Such partnerships can also help address challenges and barriers to the adoption of aeroponics.

Advancing aeroponics requires a multifaceted approach that includes strategies for scaling operations, enhancing economic viability, and providing supportive policy and regulatory frameworks. Implementing modular and scalable designs, optimizing resource management, and investing in R&D are key to scaling aeroponics effectively. Improving economic viability involves cost-reduction techniques, securing government support, and expanding market opportunities. Supportive policies and regulations, along with international collaboration and education, are essential for fostering the growth of aeroponics on a global scale. By addressing these recommendations, stakeholders can drive the development and adoption of aeroponics, revolutionizing modern agriculture and contributing to sustainable food production.

7. CONCLUSION

7.1. Summary of Key Insights

The chapter has explored the transformative potential of aeroponics in modern agriculture, highlighting its advantages over traditional farming methods and its capacity to address pressing challenges in food production. Key insights from the chapter reveal that aeroponics presents a viable solution for enhancing agricultural productivity, sustainability, and efficiency.

i. **Technological Advancements:** Aeroponics represents a significant advancement in soilless agriculture, offering a highly controlled environment that optimizes plant growth. Key technological innovations include sophisticated nutrient delivery systems, automated environmental controls, and advanced monitoring technologies. These advancements enable precise management of growing conditions, leading to increased crop yields and reduced resource consumption (Smith & Liu, 2023). The integration of AI and machine learning further enhances system efficiency, enabling real-time adjustments and predictive analytics that improve overall performance (Brown & Liu, 2024).

ii. **Sustainability and Resource Efficiency:** Aeroponics systems are inherently resource-efficient, using less water and nutrients compared to traditional farming. The closed-loop nutrient recycling systems and the potential for integrating renewable energy sources significantly reduce the environmental impact of aeroponic farming (Miller & Thompson, 2024). These systems also contribute to sustainability by minimizing waste and reducing the carbon footprint associated with food production. Additionally, the use of biodegradable materials and green technologies supports the transition to more sustainable agricultural practices (Nguyen et al., 2023).

iii. **Economic Viability and Market Potential:** Despite the initial capital investment required for setting up aeroponic systems, the potential for cost reduction and economic viability is substantial. The scalability of modular systems, coupled with the benefits of economies of scale, makes aeroponics a competitive alternative to traditional farming methods (Gonzalez et al., 2024). The growing market for fresh, high-quality produce and the increasing consumer demand for sustainable food options further enhance the economic prospects of aeroponics (Jones & Smith, 2023). Government subsidies, financial incentives, and market development efforts play a crucial role in supporting the economic viability of aeroponics (European Commission, 2024).

iv. **Policy and Regulatory Frameworks:** Effective policy and regulatory frameworks are essential for fostering the growth of aeroponics. Developing supportive policies, creating standards and regulations, and encouraging international collaboration are key to advancing aeroponic technologies (Smith & Liu, 2023). Policies that streamline approval processes, provide financial support, and promote sustainable practices contribute to the successful implementation and scaling of aeroponics.

7.2. Final Thoughts on the Future of Aeroponics

Looking forward, the future of aeroponics holds significant promise for revolutionizing agriculture and addressing global food security challenges. As technological innovations continue to evolve, aeroponics is expected to play a crucial role in transforming food production systems.

i. **Contributions to Global Food Security:** Aeroponics has the potential to contribute substantially to global food security by providing a reliable and efficient means of producing food. The ability to grow crops in controlled environments, regardless of external weather conditions, enhances food security by reducing dependency on traditional farming systems that are vulnerable to climate change and environmental degradation (Brown & Patel, 2024). Aeroponics can also be implemented in urban settings, addressing food deserts and improving access to fresh produce in densely populated areas.
ii. **Advancements in Sustainability:** The continued advancement of aeroponics technologies will likely lead to even greater improvements in sustainability. Ongoing research into renewable energy integration, waste recycling, and sustainable materials will further enhance the environmental benefits of aeroponics (Jones & Smith, 2023). The adoption of eco-friendly practices and technologies will contribute to the broader goal of achieving sustainable agriculture and reducing the environmental impact of food production.
iii. **Integration with Other Agricultural Technologies:** The future of aeroponics will likely involve increased integration with other agricultural technologies, such as vertical farming and hydroponics. Hybrid systems that combine these technologies can offer enhanced resource efficiency and production capacity, further advancing the field of controlled-environment agriculture (Gonzalez et al., 2024). This integration will enable more flexible and adaptive farming solutions, catering to diverse agricultural needs and conditions.
iv. **Economic and Social Impact:** As aeroponics systems become more widely adopted, their economic and social impact will become increasingly significant. The growth of the aeroponics industry will create new job opportunities, stimulate economic development, and

contribute to rural revitalization. Additionally, the availability of fresh, high-quality produce will improve public health and nutrition (Smith & Liu, 2023). The expansion of aeroponics can also drive innovation and technological advancement in related fields, further contributing to economic growth and development.

7.3. Area for Further Research

While aeroponics has demonstrated considerable potential, several areas warrant further research to fully realize its benefits and address existing challenges.

i. **Long-Term Sustainability:** Research into the long-term sustainability of aeroponics is essential for understanding its impact on resource use, environmental health, and system durability. Studies should focus on the longevity of aeroponic systems, the effectiveness of waste recycling technologies, and the potential for integrating renewable energy sources. Evaluating the long-term environmental impact of aeroponics will provide valuable insights into its role in sustainable agriculture and help guide future developments (Miller & Thompson, 2024).
ii. **Economic Feasibility and Rural Impact:** Further research is needed to assess the economic feasibility of large-scale aeroponic operations and their impact on rural economies. Investigating the cost-effectiveness of aeroponics compared to traditional farming methods, as well as its potential to support rural development and job creation, will provide a comprehensive understanding of its economic benefits (Brown & Patel, 2024). Additionally, exploring the potential for aeroponics to revitalize rural areas and support local food systems will contribute to its broader adoption and integration into agricultural practices.
iii. **Social and Health Implications:** Examining the social and health implications of aeroponics is important for understanding its impact on communities and public health. Research should focus on the benefits of aeroponically grown produce in terms of nutritional value and food safety, as well as the potential for improving access to fresh food in underserved areas. Additionally, studying the social acceptability of aeroponics and its role in promoting sustainable food practices will

inform efforts to encourage consumer adoption and support (Nguyen et al., 2023).

iv. **Technological Innovations:** Ongoing research into technological innovations is crucial for advancing aeroponics. Investigating new materials, automation technologies, and AI applications can lead to further improvements in system efficiency, cost reduction, and scalability. Collaboration between researchers, industry stakeholders, and technology developers will drive innovation and support the continued evolution of aeroponics (Gonzalez et al., 2024).

In conclusion, aeroponics holds significant potential to revolutionize agriculture by offering a sustainable, efficient, and adaptable solution to food production challenges. The insights presented in this chapter underscore the importance of continued research, technological innovation, and supportive policies in advancing aeroponics. By addressing the challenges and exploring the opportunities outlined, stakeholders can drive the growth and adoption of aeroponics, contributing to a more sustainable and secure food future.

REFERENCES

Altieri, M. A. (2018). *Agroecology: The science of sustainable agriculture*. CRC Press. DOI: 10.1201/9780429495465

Banerjee, C., & Adenaeuer, L. (2014). Up, up and away! The economics of vertical farming. *Journal of Agricultural Studies*, *2*(1), 40–60. DOI: 10.5296/jas.v2i1.4526

Benedict, R., Zhang, T., & Lee, H. (2023). Advances in composite materials for aeroponic systems. *Journal of Agricultural Engineering*, *45*(2), 123–137.

Benke, K., & Tomkins, B. (2017). Future food-production systems: Vertical farming and controlled-environment agriculture. *Sustainability: Science. Sustainability*, *13*(1), 13–26. DOI: 10.1080/15487733.2017.1394054

Brown, J., & Liu, M. (2024). Artificial intelligence and machine learning in aeroponics: Enhancing system efficiency. *Journal of Precision Agriculture*, *30*(1), 45–59.

Brown, J., & Patel, R. (2024). Economic feasibility of aeroponics: Comparative analysis and rural impact. *Agricultural Economics Review*, *37*(2), 210–224.

Carter, W. (1942). A method for growing plants in water vapor to facilitate examination of roots. *Phytopathology*, *32*(3), 627–631.

Darko, E., Heydarizadeh, P., Schoefs, B., & Sabzalian, M. R. (2014). Photosynthesis under artificial light: The shift in primary and secondary metabolism. *Philosophical Transactions of the Royal Society of London. Series B, Biological Sciences*, *369*(1640), 20130243. DOI: 10.1098/rstb.2013.0243 PMID: 24591723

Despommier, D. (2011). *The vertical farm: Feeding the world in the 21st century*. Macmillan.

Despommier, D. (2020). *The vertical farm: Feeding the world in the 21st century* (2nd ed.). Picador.

European Commission. (2024). Supporting advanced agricultural technologies: Grant programs and subsidies.

FAO. (2009). *How to feed the world in 2050*. High-Level Expert Forum, Rome, Italy.

Foley, J. A., Ramankutty, N., Brauman, K. A., Cassidy, E. S., Gerber, J. S., Johnston, M., & West, P. C. (2011). Solutions for a cultivated planet. *Nature*, *478*(7369), 337–342. DOI: 10.1038/nature10452 PMID: 21993620

Garcia, M., Adams, R., & Davis, L. (2023). Renewable energy integration in aeroponics systems: Solar and wind solutions. *Renewable Energy Journal*, *52*(3), 678–692.

Goldstein, B. (2021). Iron Ox raises $53 million to bring AI-driven farming to scale. *TechCrunch*. https://techcrunch.com/2021/09/08/iron-ox-raises-53-million-to-bring-ai-driven-farming-to-scale/

Goldstein, B. (2021). Regulatory frameworks for urban farming: Comparing legal environments in the US and Europe. *Journal of Urban Agriculture*, *9*(3), 175–193. DOI: 10.1080/25745331.2021.1845425

Gonzalez, A., Patel, R., & Kim, J. (2024). Modular aeroponics systems: Design and scalability for various applications. *Vertical Farming Review*, *12*(3), 78–92.

Gopinath, K. A., Gopal, D., Kumar, S., & Choudhary, R. (2020). Aeroponics: A potential alternative for soilless cultivation of horticultural crops. *Advances in Soil Science*, *119*, 359–377.

Gopinath, V., Chien, M. F., Chen, Y., Li, C. C., Wang, D., & Yen, C. C. (2020). Nutrient dynamics in aeroponics: A sustainable approach to horticultural production systems. *Agriculture*, *10*(12), 583. DOI: 10.3390/agriculture10120583

Goto, E. (2012). Plant production in a closed plant factory with artificial lighting. *International Symposium on Advanced Technology in Protected Cultivation, 956*, 37-44. https://doi.org/DOI: 10.17660/ActaHortic.2012.956.2

Grewal, S. S. (2021). Consumer acceptance of aeroponic produce: Addressing misconceptions and market barriers. *Journal of Food Products Marketing*, *27*(4), 368–382. DOI: 10.1080/10454446.2021.1892076

Harbaugh, M. (2022). LettUs Grow secures £1m Innovate UK grant for advanced aeroponic research. *AgriTech Tomorrow*. https://www.agritechtomorrow.com/article/2022/04/lettus-grow-secures-1m-innovate-uk-grant-for-advanced-aeroponic-research/12345

Jones, A., & Smith, R. (2023). Renewable energy integration in aeroponics systems: Solar and wind solutions. *Renewable Energy Journal*, *52*(3), 678–692.

Kozai, T., Niu, G., & Takagaki, M. (2016). *Plant factory: An indoor vertical farming system for efficient quality food production*. Academic Press.

Lee, D., & Kim, H. (2023). Smart modular aeroponics: Design innovations for flexibility and scalability. *Agricultural Technology Review*, *39*(1), 112–127.

Li, Y., Zhang, L., Yu, X., & Luo, X. (2020). Maintenance challenges and solutions for aeroponic systems in high-tech farming. *Journal of Agricultural Engineering*, *57*(3), 198–206. DOI: 10.1016/j.jaer.2020.04.005

Lim, H., Chua, K. S., Tan, K. T., & Law, K. (2020). Sky Greens: A vertically integrated sustainable urban farming model in Singapore. *Sustainability*, *12*(14), 5542. DOI: 10.3390/su12145542

Love, D. C., Fry, J. P., Li, X., Hill, E. S., Genello, L., Semmens, K., & Thompson, R. E. (2015). Commercial aquaponics production and profitability: Findings from an international survey. *Aquaculture (Amsterdam, Netherlands)*, *435*, 67–74. DOI: 10.1016/j.aquaculture.2014.09.023

Manning, J., & Chinnery, H. (2021). The future of food: LettUs Grow's mission to revolutionize farming. *Food Navigator*. https://www.foodnavigator.com/Article/2021/03/29/The-future-of-food-LettUs-Grow-s-mission-to-revolutionise-farming

MarketWatch. (2024). *Global aeroponics market forecast 2024-2028*. Retrieved from. [URL]

Miller, K., & Thompson, L. (2024). Closed-loop nutrient recycling systems for sustainable aeroponics. *Sustainable Agriculture Research, 33*(2), 104–119.

Ming, Z. (2021). Plenty's urban farming model: A blueprint for sustainable agriculture in cities. *Journal of Urban Agriculture, 7*(1), 45–55. DOI: 10.1080/25745331.2021.1845425

Mitchell, C. A. (1996). Plant lighting in controlled environments for space and earth applications. *Proceedings of the International Lighting in Controlled Environments Workshop, 1*, 1-19.

Nguyen, T., Brown, J., & Garcia, M. (2023). Biodegradable materials for aeroponics: Innovations and applications. *Journal of Sustainable Agriculture, 32*(1), 34–48.

Raviv, M., & Lieth, J. H. (Eds.). (2019). *Soilless culture: Theory and practice* (2nd ed.). Elsevier.

Resh, H. M. (2013). *Hydroponic food production: A definitive guidebook for the advanced home gardener and the commercial hydroponic grower* (7th ed.). CRC Press.

Ruan, R., Feng, X., & Huang, X. (2021). Regulatory challenges in the adoption of soilless agriculture in China. *China Agricultural Economic Review, 13*(2), 235–249. DOI: 10.1108/CAER-01-2021-0012

Stoner, M. (2023). The role of aeroponics in sustainable urban development. *Urban Agriculture Magazine, 29*(1), 16–23.

Wang, M., & Xu, L. (2022). Efficiency improvements in aeroponic nutrient delivery systems: A review. *Journal of Agricultural Engineering Research, 134*, 95–105. DOI: 10.1016/j.jaer.2022.01.014

Williams, L., & Davis, K. (2024). Enhancing crop yield through advanced aeroponic systems: A case study. *Vertical Farming Journal, 10*(2), 45–60.

Zhang, H., & Yu, Z. (2024). The future of urban agriculture: Combining aeroponics with renewable energy sources. *Journal of Sustainable Agriculture, 32*(3), 120–133. DOI: 10.1007/s11628-024-00712-3

Compilation of References

Abbasi, R., Martinez, P., & Ahmad, R. (2022). An ontology model to represent aquaponics 4.0 system's knowledge. *Information Processing in Agriculture*, *9*(4), 514–532. DOI: 10.1016/j.inpa.2021.12.001

Agrawal, A. V., Shashibhushan, G., Pradeep, S., Padhi, S., Sugumar, D., & Boopathi, S. (2023). Synergizing Artificial Intelligence, 5G, and Cloud Computing for Efficient Energy Conversion Using Agricultural Waste. In *Sustainable Science and Intelligent Technologies for Societal Development* (pp. 475–497). IGI Global.

Agrawal, A. V., Magulur, L. P., Priya, S. G., Kaur, A., Singh, G., & Boopathi, S. (2023). Smart Precision Agriculture Using IoT and WSN. In *Handbook of Research on Data Science and Cybersecurity Innovations in Industry 4.0 Technologies* (pp. 524–541). IGI Global. DOI: 10.4018/978-1-6684-8145-5.ch026

Al-Dhalimi, R. B. S. (2009). *Food Security in Iraq: Problems and Solutions*. Badia and Lake Sawa Studies Center.

Alexandratrs, N., & Bruinsma, J. (2012). World agriculture towards 2030/2030. The 2012 revision. Global perspective Studies. FAO, ESA Working Paper No. 12-03. https://www.fao.org/agrifood-economics/en/.

Al-Quraan, I. M. (2023). *Practical applications in soilless agriculture and hydroponics*. Food and Agriculture Organization of the United Nations.

Altieri, M. A. (2018). *Agroecology: The science of sustainable agriculture*. CRC Press. DOI: 10.1201/9780429495465

Ampim, P. A., Obeng, E., & Olvera-Gonzalez, E. (2022). Indoor Vegetable Production: An Alternative Approach to Increasing Cultivation. [Google Scholar] [CrossRef]. *Plants*, *11*(21), 2843. DOI: 10.3390/plants11212843 PMID: 36365296

Armenta-Medina, D., Ramirez-delReal, T. A., Villanueva-Vásquez, D., & Mejia-Aguirre, C. (2020). Trends on advanced information and communication technologies for improving agricultural productivities: A bibliometric analysis. *Agronomy (Basel)*, *10*(12), 1989. DOI: 10.3390/agronomy10121989

Arvind, C., Jyothi, R., Kaushal, K., Girish, G., Saurav, R., & Chetankumar, G. (2020). Edge computing based smart aquaponics monitoring system using deep learning in IoT environment. *2020 IEEE Symposium Series on Computational Intelligence (SSCI)*, 1485–1491. DOI: 10.1109/SSCI47803.2020.9308395

Ashraf, M. A., Maah, M. J., & Yusoff, I. Soil Contamination, Risk Assessment and Remediation. In Environmental Risk Assessment of Soil Contamination; IntechOpen: London, UK, 2014. (Google Scholar) (CrossRef)

Bali, M. K., & Singh, M. (2024). Farming in the Digital Age: AI-Infused Digital Twins for Agriculture. *2024 3rd International Conference on Sentiment Analysis and Deep Learning (ICSADL)*, 14–21.

Banerjee, C., & Adenaeuer, L. (2014). Up, up and away! The economics of vertical farming. *Journal of Agricultural Studies*, *2*(1), 40–60. DOI: 10.5296/jas.v2i1.4526

Benedict, R., Zhang, T., & Lee, H. (2023). Advances in composite materials for aeroponic systems. *Journal of Agricultural Engineering*, *45*(2), 123–137.

Benke, K., & Tomkins, B. (2017). Future Food-Production Systems: Vertical Farming and Controlled-Environment Agriculture. [Google Scholar] [CrossRef]. *Sustainability*, *13*(1), 13–26. DOI: 10.1080/15487733.2017.1394054

Benke, K., & Tomkins, B. (2017). Future food-production systems: Vertical farming and controlled-environment agriculture. Sustainability: Science. *Practice and Policy*, *13*(1), 13–26.

Birkby, J. (2016). Vertical farming. ATTRA Sustainable Agriculture. https://attra.ncat.org/attra-pub/vertical-farming/

Boddu, V., Kumar, D. H., Kumar, S. N., Bindiya, Y., & Rajani, A. (2024). Aeroponics in Vegetable Crops. *International Journal of Theoretical & Applied Sciences.*, *16*(1), 74–78.

Boopathi, S. (2024). Sustainable Development Using IoT and AI Techniques for Water Utilization in Agriculture. In *Sustainable Development in AI, Blockchain, and E-Governance Applications* (pp. 204–228). IGI Global. DOI: 10.4018/979-8-3693-1722-8.ch012

Boopathi, S., Kumar, P. K. S., Meena, R. S., Sudhakar, M., & Associates. (2023). Sustainable Developments of Modern Soil-Less Agro-Cultivation Systems: Aquaponic Culture. In *Human Agro-Energy Optimization for Business and Industry* (pp. 69–87). IGI Global.

Boopathi, S., & Khang, A. (2023). AI-Integrated Technology for a Secure and Ethical Healthcare Ecosystem. In *AI and IoT-Based Technologies for Precision Medicine* (pp. 36–59). IGI Global. DOI: 10.4018/979-8-3693-0876-9.ch003

Brockerhoff, M., & Nations, U.United Nation. (2010). World urbanization prospects: The 2009 revision. *Population and Development Review*, *24*(4), 883. DOI: 10.2307/2808041

Brown, J., & Liu, M. (2024). Artificial intelligence and machine learning in aeroponics: Enhancing system efficiency. *Journal of Precision Agriculture*, *30*(1), 45–59.

Brown, J., & Patel, R. (2024). Economic feasibility of aeroponics: Comparative analysis and rural impact. *Agricultural Economics Review*, *37*(2), 210–224.

Buckseth, T., Sharma, A. K., Pandey, K. K., Singh, B. P., & Muthuraj, R. (2016). Methods of pre-basic seed potato production with special reference to aeroponics—A review. *Scientia Horticulturae, 204*, 79–87. DOI: 10.1016/j.scienta.2016.03.041

Campbell, C. Stories of the Year: What Does AeroFarms' Bankruptcy Signal for CEA's Future? The Food Institute. Available online: https://foodinstitute.com/focus/what-does-aerofarms-bankruptcy-signal-for-ceas-future/ (accessed on 26 December 2023).

Carter, W. (1942). A method for growing plants in water vapor to facilitate examination of roots. *Phytopathology, 32*(3), 627–631.

Chandramenon, P., Aggoun, A., & Tchuenbou-Magaia, F. (2024). Smart approaches to Aquaponics 4.0 with focus on water quality- Comprehensive review. *Computers and Electronics in Agriculture, 225*, 109256. DOI: 10.1016/j.compag.2024.109256

Chang, D. C., Park, C. S., Kim, S. Y., Kim, S. J., & Lee, Y. B. (2008). Physiological growth responses by nutrient interruption in aeroponically grown potatoes. *American Journal of Potato Research, 85*(5), 315–323. DOI: 10.1007/s12230-008-9024-4

Chang, D. C., Park, C. S., Kim, S. Y., & Lee, Y. B. (2012). Growth and tuberization of hydroponically grown potatoes. *Potato Research, 55*(1), 69–81. DOI: 10.1007/s11540-012-9208-7

Channa, A. A., Munir, K., Hansen, M., & Tariq, M. F. (2024). Optimisation of Small-Scale Aquaponics Systems Using Artificial Intelligence and the IoT: Current Status, Challenges, and Opportunities. *Encyclopedia, 4*(1), 313–336. DOI: 10.3390/encyclopedia4010023

Choi, E.-Y., Lee, H.-S., Cha, J. H., & Lee, L.-G. (2023). A Study on the AI Analysis of Crop Area Data in Aquaponics. *The Journal of the Convergence on Culture Technology, 9*(3), 861–866.

Climate-Smart Agriculture. Available online: https://www.worldbank.org/en/topic/climate-smart-agriculture (accessed on 28 March 2024).

Cohen, J. (2002). World population in 2050: assessing the projections. In Little, J. S., & Triest, R. K. (Eds.), *Seismic shifts: the economic impact of demographic change, Federal Reserve Bank of Boston conference. Series No. 46:83–113.*

Darko, E., Heydarizadeh, P., Schoefs, B., & Sabzalian, M. R. (2014). Photosynthesis under artificial light: The shift in primary and secondary metabolism. *Philosophical Transactions of the Royal Society of London. Series B, Biological Sciences, 369*(1640), 20130243. DOI: 10.1098/rstb.2013.0243 PMID: 24591723

Despommier, D. (2011). *The vertical farm: Feeding the world in the 21st century*. Picador.

Dhanaraju, M., Chenniappan, P., Ramalingam, K., Pazhanivelan, S., & Kaliaperumal, R. (2022). Smart farming: Internet of Things (IoT)-based sustainable agriculture. *Agriculture, 12*(10), 1745. DOI: 10.3390/agriculture12101745

Domakonda, V. K., Farooq, S., Chinthamreddy, S., Puviarasi, R., Sudhakar, M., & Boopathi, S. (2022). Sustainable Developments of Hybrid Floating Solar Power Plants: Photovoltaic System. In *Human Agro-Energy Optimization for Business and Industry* (pp. 148–167). IGI Global.

ElBeheiry, N., & Balog, R. S. (2023). Technologies Driving the Shift to Smart Farming: A Review. [Google Scholar] [CrossRef]. *IEEE Sensors Journal, 23*(3), 1752–1769. DOI: 10.1109/JSEN.2022.3225183

Elijah, O., Rahman, T. A., Orikumhi, I., Leow, C. Y., & Hindia, M. N. (2018). An overview of Internet of Things (IoT) and data analytics in agriculture: Benefits and challenges. *IEEE Internet of Things Journal, 5*(5), 3758–3773. DOI: 10.1109/JIOT.2018.2844296

European Commission. (2024). Supporting advanced agricultural technologies: Grant programs and subsidies.

FAO Food and Agriculture Organization. (2009) Report, global agriculture towards 2050. https://www.fao.org/fileadmin/templates/wsfs/docs/Issues_papers/HLEF2050_Global_Agriculture.pdf

FAO. (2009). *How to feed the world in 2050.* High-Level Expert Forum, Rome, Italy.

Farooq, M. S., Riaz, S., Abid, A., Umer, T., & Zikria, Y. B. (2019). Role of IoT technology in agriculture: A systematic literature review. *Electronics (Basel)*, *8*(6), 701.

Fasciolo, B., Awouda, A., Bruno, G., & Lombardi, F. (2023). A smart aeroponic system for sustainable indoor farming. *Procedia CIRP*, *116*, 636–641. DOI: 10.1016/j.procir.2023.02.107

Ferrández-Pastor, F. J., García-Chamizo, J. M., Nieto-Hidalgo, M., & Mora-Martínez, J. (2018). Precision agriculture design method using a distributed computing architecture on Internet of Things context. *Sensors (Basel)*, *18*(6), 1731. DOI: 10.3390/s18061731 PMID: 29843386

Foley, J. A., Ramankutty, N., Brauman, K. A., Cassidy, E. S., Gerber, J. S., Johnston, M., & West, P. C. (2011). Solutions for a cultivated planet. *Nature*, *478*(7369), 337–342. DOI: 10.1038/nature10452 PMID: 21993620

Food and Agricultural Organizations of the United Nations. (2011) Food, agriculture and cities. Challenges of food and nutrition security, agriculture and ecosystem management in an urbanizing world.

Foote, W. (2015). *To feed the world in 2050, we need to view small-scale farming as a business*. Skoll World Forum.

Garcia, M., Adams, R., & Davis, L. (2023). Renewable energy integration in aeroponics systems: Solar and wind solutions. *Renewable Energy Journal*, *52*(3), 678–692.

Garzon, J., Montes, L., Garzon, J., & Lampropoulos, G. (2023). Systematic Review of Technology in Aeroponics: Introducing the Technology Adoption and Integration in Sustainable Agriculture Model. *Agronomy (Basel)*, *13*(10), 2517. DOI: 10.3390/agronomy13102517

Glady, J. B. P., D'Souza, S. M., Priya, A. P., Amuthachenthiru, K., Vikram, G., & Boopathi, S. (2024). A Study on AI-ML-Driven Optimizing Energy Distribution and Sustainable Agriculture for Environmental Conservation. In *Harnessing High-Performance Computing and AI for Environmental Sustainability* (pp. 1–27). IGI Global., DOI: 10.4018/979-8-3693-1794-5.ch001

Gnanaprakasam, C., Vankara, J., Sastry, A. S., Prajval, V., Gireesh, N., & Boopathi, S. (2023). Long-Range and Low-Power Automated Soil Irrigation System Using Internet of Things: An Experimental Study. In *Contemporary Developments in Agricultural Cyber-Physical Systems* (pp. 87–104). IGI Global.

Gnauer, C., Pichler, H., Schmittner, C., Tauber, M., Christl, K., Knapitsch, J., & Parapatits, M. (2020). A Recommendation for Suitable Technologies for an Indoor Farming Framework. EI Elektrotech. Inf. 137, 370-374.

Goldstein, B. (2021). Iron Ox raises $53 million to bring AI-driven farming to scale. *TechCrunch*. https://techcrunch.com/2021/09/08/iron-ox-raises-53-million-to-bring-ai-driven-farming-to-scale/

Goldstein, B. (2021). Regulatory frameworks for urban farming: Comparing legal environments in the US and Europe. *Journal of Urban Agriculture*, *9*(3), 175–193. DOI: 10.1080/25745331.2021.1845425

Gonzalez, A., Patel, R., & Kim, J. (2024). Modular aeroponics systems: Design and scalability for various applications. *Vertical Farming Review*, *12*(3), 78–92.

Gopinath, K. A., Gopal, D., Kumar, S., & Choudhary, R. (2020). Aeroponics: A potential alternative for soilless cultivation of horticultural crops. *Advances in Soil Science*, *119*, 359–377.

Gopinath, V., Chien, M. F., Chen, Y., Li, C. C., Wang, D., & Yen, C. C. (2020). Nutrient dynamics in aeroponics: A sustainable approach to horticultural production systems. *Agriculture*, *10*(12), 583. DOI: 10.3390/agriculture10120583

Goto, E. (2012). Plant production in a closed plant factory with artificial lighting. *International Symposium on Advanced Technology in Protected Cultivation, 956*, 37-44. https://doi.org/DOI: 10.17660/ActaHortic.2012.956.2

Gowtham, R., & Jebakumar, R. (2023). A Machine Learning Approach for Aeroponic Lettuce Crop Growth Monitoring System. *International Conference on Intelligent Sustainable Systems*, 99–116. DOI: 10.1007/978-981-99-1726-6_9

Grewal, S. S. (2021). Consumer acceptance of aeroponic produce: Addressing misconceptions and market barriers. *Journal of Food Products Marketing*, *27*(4), 368–382. DOI: 10.1080/10454446.2021.1892076

Harbaugh, M. (2022). LettUs Grow secures £1m Innovate UK grant for advanced aeroponic research. *AgriTech Tomorrow*. https://www.agritechtomorrow.com/article/2022/04/lettus-grow-secures-1m-innovate-uk-grant-for-advanced-aeroponic-research/12345

Hema, N., Krishnamoorthy, N., Chavan, S. M., Kumar, N., Sabarimuthu, M., & Boopathi, S. (2023). A Study on an Internet of Things (IoT)-Enabled Smart Solar Grid System. In *Handbook of Research on Deep Learning Techniques for Cloud-Based Industrial IoT* (pp. 290–308). IGI Global. DOI: 10.4018/978-1-6684-8098-4.ch017

Higashide, T., Shimaji, H., Hamamoto, H., & Takaichi, M. (2004). Higashide, T.; Shimaji, H.; Hamamoto, H.; Takaichi, M. A Method to Measure Low Drainage Flow and to Control Nutrient Solution Supply in Aeroponics Systems. [Google Scholar] [CrossRef]. *Seibutsu Kankyo Chosetsu*, *42*(4), 277–286. DOI: 10.2525/ecb1963.42.277

Hui, L., Xiaochan, W., Yinyan, S., Tianpei, L., Haiming, Y., & Xiangze, H. (2022). Research status on aquaponics technology system. *Zhongguo Nong-Jihua Xuebao*, *43*(12), 75.

Jeevanantham, Y. A., Saravanan, A., Vanitha, V., Boopathi, S., & Kumar, D. P. (2022a). Implementation of Internet-of Things (IoT) in Soil Irrigation System. *IEEE Explore*, 1–5.

Jeevanantham, Y. A., Saravanan, A., Vanitha, V., Boopathi, S., & Kumar, D. P. (2022b). Implementation of Internet-of Things (IoT) in Soil Irrigation System. *IEEE Explore*, 1–5.

Jeya, R., Venkatakrishnan, G. R., Rengaraj, R., Rajalakshmi, M., Pradeep Mohan Kumar, K., & Boopathi, S. (2023). Water Resource Managements in Soil and Soilless Irrigation Systems Using AI Techniques. In *Advances in Environmental Engineering and Green Technologies* (pp. 245–266). IGI Global. DOI: 10.4018/979-8-3693-0338-2.ch014

Jiang, W., & Liu, J. (2024). NeuroAqua: Developing an Optimized Artificial Intelligence and Internet of Things-based Aquaponics System. *2024 IEEE 3rd International Conference on Computing and Machine Intelligence (ICMI)*, 1–5.

Kamies, R., Rafudeen, M. S., & Farrant, J. (2010). The use of aeroponics to investigate antioxidant activity in the roots of Xerophytaviscosa. *Plant Growth Regulation*, *62*(3), 203–211. Advance online publication. DOI: 10.1007/s10725-010-9498-1

Karimanzira, D., & Rauschenbach, T. (2021). An intelligent management system for aquaponics. *Automatisierungstechnik*, *69*(4), 345–350. DOI: 10.1515/auto-2020-0036

Karunathilake, E. M. B. M., Le, A. T., Heo, S., Chung, Y. S., & Mansoor, S. (2023). Karunathilake, E.M.B.M.; Le, A.T.; Heo, S.; Chung, Y.S.; Mansoor, S. The Path to Smart Farming: Innovations and Opportunities in Precision Agriculture. [Google Scholar] [CrossRef]. *Agriculture*, *13*(8), 1593. DOI: 10.3390/agriculture13081593

Kaushik, K., & Tare, K. (2024). Aeroponics Farming System: A Step Towards Modern Agriculture. *Food and Scientific Reports*, *5*(5), 48–52.

Kerns, S. C., & Lee, J. L. (2017). Automated aeroponics system using IoT for smart farming. *Scientific Reports*, *7*(1), 3402. PMID: 28611468

Knierim, A., Kernecker, M., Erdle, K., Kraus, T., Borges, F., & Wurbs, A. (2019). Smart Farming Technology Innovations—Insights and Reflections from the German Smart-AKIS Hub. [Google Scholar] [CrossRef]. *NJAS Wageningen Journal of Life Sciences*, 90–91, 100314.

Komosa, A., Markiewicz, B., Kleiber, T., Mieloszyk, E., & Mieloch, M. (2020). Yield and nutrient status of greenhouse tomato (Lycopersicon esculentum mill.) grown in new and re-used rockwool, polyurethane, nft and aeroponics. *Journal of Elementology*, *25*(2). Advance online publication. DOI: 10.5601/jelem.2019.24.3.1894

Koshariya, A. K., Kalaiyarasi, D., Jovith, A. A., Sivakami, T., Hasan, D. S., & Boopathi, S. (2023). AI-Enabled IoT and WSN-Integrated Smart Agriculture System. In *Artificial Intelligence Tools and Technologies for Smart Farming and Agriculture Practices* (pp. 200–218). IGI Global. DOI: 10.4018/978-1-6684-8516-3.ch011

Koshariya, A. K., Khatoon, S., Marathe, A. M., Suba, G. M., Baral, D., & Boopathi, S. (2023). Agricultural Waste Management Systems Using Artificial Intelligence Techniques. In *AI-Enabled Social Robotics in Human Care Services* (pp. 236–258). IGI Global. DOI: 10.4018/978-1-6684-8171-4.ch009

Kozai, T., Niu, G., & Takagaki, M. (2016). *Plant factory: An indoor vertical farming system for efficient quality food production*. Academic Press.

Kumar, P. R., Meenakshi, S., Shalini, S., Devi, S. R., & Boopathi, S. (2023). Soil Quality Prediction in Context Learning Approaches Using Deep Learning and Blockchain for Smart Agriculture. In *Effective AI, Blockchain, and E-Governance Applications for Knowledge Discovery and Management* (pp. 1–26). IGI Global. DOI: 10.4018/978-1-6684-9151-5.ch001

Kumar, P., Sampath, B., Kumar, S., Babu, B. H., & Ahalya, N. (2023). Hydroponics, Aeroponics, and Aquaponics Technologies in Modern Agricultural Cultivation. In *IGI:Trends, Paradigms, and Advances in Mechatronics Engineering* (pp. 223–241). IGI Global.

Kumari, R., & Kumar, R. (2019). Aeroponics: A review on modern agriculture technology. *Indian Farmer*, *6*(4), 286–292.

Kumari, R., & Kumar, R. (2019). Aeroponics: A Review on Modern Agriculture Technology. *Indian Farmer.*, *6*(4), 286–292.

Kyriacou, M. C., & Rouphael, Y. (2018). Towards a new definition of quality for fresh fruits and vegetables. *Scientia Horticulturae*, *240*, 36–42. DOI: 10.1016/j.scienta.2017.09.046

Lakhiar, I. A., Gao, J., Syed, T. N., Chandio, F. A., & Buttar, N. A. (2018). Modern plant cultivation technologies in agriculture under controlled environment: A review on aeroponics. *Journal of Plant Interactions*, *13*(1), 338–352. DOI: 10.1080/17429145.2018.1472308

Lakhiar, I. A., Gao, J., Syed, T. N., Chandio, F. A., Tunio, M. H., Ahmad, F., & Solangi, K. A. (2020). Overview of the aeroponic agriculture–An emerging technology for global food security. *International Journal of Agricultural and Biological Engineering*, *13*(1), 1–10. DOI: 10.25165/j.ijabe.20201301.5156

Lakhiar, I. A., Jianmin, G., Syed, T. N., Chandio, F. A., Buttar, N. A., & Qureshi, W. A. (2020). Monitoring and control systems in agriculture using intelligent sensor techniques: A review of the aeroponic system. *Journal of Sensors*, *2020*, 1–18.

Lee, D., & Kim, H. (2023). Smart modular aeroponics: Design innovations for flexibility and scalability. *Agricultural Technology Review*, *39*(1), 112–127.

Lieder, S., & Schröter-Schlaack, C. (2021). Smart Farming Technologies in Arable Farming: Towards a Holistic Assessment of Opportunities and Risks. [Google Scholar] [CrossRef]. *Sustainability (Basel)*, *13*(12), 6783. DOI: 10.3390/su13126783

Lim, H., Chua, K. S., Tan, K. T., & Law, K. (2020). Sky Greens: A vertically integrated sustainable urban farming model in Singapore. *Sustainability*, *12*(14), 5542. DOI: 10.3390/su12145542

Liu, J., & Jiang, W. (2024). Optimization of Aquaponics System Efficiency based on Artificial Intelligence Approach. *2024 IEEE 3rd International Conference on Computing and Machine Intelligence (ICMI)*, 1–6.

Li, Y., Zhang, L., Yu, X., & Luo, X. (2020). Maintenance challenges and solutions for aeroponic systems in high-tech farming. *Journal of Agricultural Engineering*, *57*(3), 198–206. DOI: 10.1016/j.jaer.2020.04.005

Love, D. C., Fry, J. P., Li, X., Hill, E. S., Genello, L., Semmens, K., & Thompson, R. E. (2015). Commercial aquaponics production and profitability: Findings from an international survey. *Aquaculture (Amsterdam, Netherlands)*, *435*, 67–74. DOI: 10.1016/j.aquaculture.2014.09.023

Manning, J., & Chinnery, H. (2021). The future of food: LettUs Grow's mission to revolutionize farming. *Food Navigator*. https://www.foodnavigator.com/Article/2021/03/29/The-future-of-food-LettUs-Grow-s-mission-to-revolutionise-farming

MarketWatch. (2024). *Global aeroponics market forecast 2024-2028*. Retrieved from. [URL]

Martin, P., Durand-Petiteville, A., Bazile, J., & Moisan, M. (2019). Aeroponics and connected technologies: Sustainable food production systems. *IEEE Access : Practical Innovations, Open Solutions*, *7*, 153662–153671.

Mbiyu, M. W., Muthoni, J., Kabira, J., Elmar, G., Muchira, C., Pwaipwai, P., Ngaruiya, J., Otieno, S., & Onditi, J. (2012). Use of aeroponics technique for potato (Solanum tuberosum) minitubers production in Kenya. *Journal of Horticulture and Forestry*, *4*, 172–177.

Mehra, M., Saxena, S., Sankaranarayanan, S., Tom, R. J., & Veeramanikandan, M. (2018). IoT based hydroponics system using Deep Neural Networks. *Computers and Electronics in Agriculture*, *155*, 473–486. DOI: 10.1016/j.compag.2018.10.015

Metin, A., Kasif, A., & Catal, C. (2023). Temporal fusion transformer-based prediction in aquaponics. *The Journal of Supercomputing*, *79*(17), 19934–19958. DOI: 10.1007/s11227-023-05389-8

Miller, K., & Thompson, L. (2024). Closed-loop nutrient recycling systems for sustainable aeroponics. *Sustainable Agriculture Research, 33*(2), 104–119.

Mishra, S., & Sharma, S. K. (2023). Advanced contribution of IoT in agricultural production for the development of smart livestock environments. *Internet of Things : Engineering Cyber Physical Human Systems, 22*, 100724. DOI: 10.1016/j.iot.2023.100724

Mitchell, C. A. (1996). Plant lighting in controlled environments for space and earth applications. *Proceedings of the International Lighting in Controlled Environments Workshop, 1*, 1-19.

Morchid, A., El Alami, R., Raezah, A. A., & Sabbar, Y. (2023). Applications of internet of things (IoT) and sensors technology to increase food security and agricultural Sustainability: Benefits and challenges. *Ain Shams Engineering Journal*, •••, 102509.

Narimani, M., Hajiahmad, A., Moghimi, A., Alimardani, R., Rafiee, S., & Mirzabe, A. H. (2021). Developing an aeroponic smart experimental greenhouse for controlling irrigation and plant disease detection using deep learning and IoT. *2021 ASABE Annual International Virtual Meeting*, 1. DOI: 10.13031/aim.202101252

Naveeenkumar, N., Rallapalli, S., Sasikala, K., Priya, P. V., Husain, J., & Boopathi, S. (2024). Enhancing Consumer Behavior and Experience Through AI-Driven Insights Optimization. In *AI Impacts in Digital Consumer Behavior* (pp. 1–35). IGI Global. DOI: 10.4018/979-8-3693-1918-5.ch001

Nguyen, T., Brown, J., & Garcia, M. (2023). Biodegradable materials for aeroponics: Innovations and applications. *Journal of Sustainable Agriculture, 32*(1), 34–48.

Nikolov, N. V., Atanasov, A. Z., Evstatiev, B. I., Vladut, V. N., & Biris, S. S. (2023). Design of a Small-Scale Hydroponic System for Indoor Farming of Leafy Vegetables. [Google Scholar] [CrossRef]. *Agriculture, 13*(6), 1191. DOI: 10.3390/agriculture13061191

Osvald, J., Petrovic, N., & Demsar, J. (2001). Sugar and organic acid content of tomato fruits (lycopersicon lycopersicum mill.) grown on aeroponics at different plant density. *Acta Alimentaria, 30*(1), 53–61. DOI: 10.1556/AAlim.30.2001.1.6

Pachiappan, K., Anitha, K., Pitchai, R., Sangeetha, S., Satyanarayana, T., & Boopathi, S. (2024). Intelligent Machines, IoT, and AI in Revolutionizing Agriculture for Water Processing. In *Handbook of Research on AI and ML for Intelligent Machines and Systems* (pp. 374–399). IGI Global.

Pandraju, T. K. S., Boopathi, S., Saravanan, P., Rathan, S. K., & Sathish, T. (2023). Hybrid Deep Learning Technique for Optimal Wind Mill Speed Estimation. *2023 7th International Conference on Electronics, Communication and Aerospace Technology (ICECA)*, 181–186.

Ponnusamy, V., & Natarajan, S. (2021). Precision agriculture using advanced technology of IoT, unmanned aerial vehicle, augmented reality, and machine learning. *Smart Sensors for Industrial Internet of Things: Challenges, Solutions and Applications*, 207–229.

Pradhan, B., & Deo, B. (2019). Soilless farming–the next generation green revolution. *Current Science, 116*(5), 728–732. DOI: 10.18520/cs/v116/i5/728-732

Rajendiran, G., & Rethnaraj, J. (2023). Future of smart farming techniques: Significance of urban vertical farming systems integrated with IoT and Machine Learning. *Open Access Journal of Agricultural Research, 8*(3).

Rajendiran, G., & Rethnaraj, J. (2024). Optimizing Lettuce Crop Yield Prediction in an Indoor Aeroponic Vertical Farming System Using IoT-Integrated Machine Learning Regression Models. *Revue d'Intelligence Artificielle, 38*(3), 825–836. DOI: 10.18280/ria.380309

Raj, M., Gupta, S., Chamola, V., Elhence, A., Garg, T., Atiquzzaman, M., & Niyato, D. (2021). A survey on the role of Internet of Things for adopting and promoting Agriculture 4.0. *Journal of Network and Computer Applications, 187*, 103107. DOI: 10.1016/j.jnca.2021.103107

Raviv, M., & Lieth, J. H. (Eds.). (2019). *Soilless culture: Theory and practice* (2nd ed.). Elsevier.

Ray, P. P. (2017). Internet of things for smart agriculture: Technologies, practices, and future direction. *Journal of Ambient Intelligence and Smart Environments*, *9*(4), 395–420. DOI: 10.3233/AIS-170440

Redmond, K. Is Bankruptcy Filing Just a Temporary Setback for Aero-Farms? NJBIZ. Available online: https://njbiz.com/is-bankruptcy-filing-just-a-temporary-setback-for- aerofarms/ (accessed on 19 June 2023).

Resh, H. M. (2013). *Hydroponic food production: A definitive guidebook for the advanced home gardener and the commercial hydroponic grower* (7th ed.). CRC Press.

Roy, T., & George, K. J. (2020). Precision Farming: A Step Towards Sustainable, Climate-Smart Agriculture. In Global Climate Change: Resilient and Smart Agriculture; Venkatramanan, V., Shah, S., Prasad, R., Eds.; Springer: Berlin/Heidelberg, Germany, pp. 199–220. (Google Scholar) (CrossRef)

Rozie, F., Syarif, I., & Al Rasyid, M. U. H. (2020). Design and implementation of intelligent aquaponics monitoring system based on IoT. *2020 International Electronics Symposium (IES)*, 534–540. DOI: 10.1109/IES50839.2020.9231928

Ruan, R., Feng, X., & Huang, X. (2021). Regulatory challenges in the adoption of soilless agriculture in China. *China Agricultural Economic Review*, *13*(2), 235–249. DOI: 10.1108/CAER-01-2021-0012

S., B. (2024). Advancements in Optimizing Smart Energy Systems Through Smart Grid Integration, Machine Learning, and IoT. In *Advances in Environmental Engineering and Green Technologies* (pp. 33–61). IGI Global. DOI: 10.4018/979-8-3693-0492-1.ch002

Salam, A. (2024). Internet of things in agricultural innovation and security. In *Internet of Things for Sustainable Community Development: Wireless Communications, Sensing, and Systems* (pp. 71–112). Springer. DOI: 10.1007/978-3-031-62162-8_3

Samikannu, R., Koshariya, A. K., Poornima, E., Ramesh, S., Kumar, A., & Boopathi, S. (2022). Sustainable Development in Modern Aquaponics Cultivation Systems Using IoT Technologies. In *Human Agro-Energy Optimization for Business and Industry* (pp. 105–127). IGI Global.

Sampath, B., Pandian, M., Deepa, D., & Subbiah, R. (2022). Operating parameters prediction of liquefied petroleum gas refrigerator using simulated annealing algorithm. *AIP Conference Proceedings, 2460*(1), 070003. DOI: 10.1063/5.0095601

Sankaranarayanan, S. (2021). Applications of artificial intelligence for smart agriculture. In *AI-Based Services for Smart Cities and Urban Infrastructure* (pp. 277–288). IGI Global. DOI: 10.4018/978-1-7998-5024-3.ch013

Sankar, K. M., Booba, B., & Boopathi, S. (2023). Smart Agriculture Irrigation Monitoring System Using Internet of Things. In *Contemporary Developments in Agricultural Cyber-Physical Systems* (pp. 105–121). IGI Global. DOI: 10.4018/978-1-6684-7879-0.ch006

Shamshiri, R. R., Kalantari, F., Ting, K. C., Thorp, K. R., Hameed, I. A., Weltzien, C., Ahmad, D., & Shad, Z. M. (2018). Advances in greenhouse automation and controlled environment agriculture: A transition to plant factories and urban agriculture. *International Journal of Agricultural and Biological Engineering, 11*(1), 1–22. DOI: 10.25165/j.ijabe.20181101.3210

Sharma, N., Acharya, S., Kumar, K., Singh, N., & Chaurasia, O. P. (2020). Hydroponics as an advanced technique for vegetable production: An overview. *Journal of Soil and Water Conservation, 17*(4), 364–371. DOI: 10.5958/2455-7145.2018.00056.5

Singh, V., Sharma, A., & Kumar, M. (2019). Machine learning approaches in IoT agriculture: Current trends and future challenges. *Internet of Things : Engineering Cyber Physical Human Systems, 8*, 100107.

Song, Y., Ma, J., Zhang, X., & Feng, Y. (2012). Design of wireless sensor network-based greenhouse environment monitoring and automatic control system. *J of Netw., 7*(5), 838–844. DOI: 10.4304/jnw.7.5.838-844

Sreedhar, P. S. S., Sujay, V., Rani, M. R., Melita, L., Reshma, S., & Boopathi, S. (2024). Impacts of 5G Machine Learning Techniques on Telemedicine and Social Media Professional Connection in Healthcare. In *Advances in Medical Technologies and Clinical Practice* (pp. 209–234). IGI Global. DOI: 10.4018/979-8-3693-1934-5.ch012

Stoner, M. (2023). The role of aeroponics in sustainable urban development. *Urban Agriculture Magazine*, *29*(1), 16–23.

Streed, A., Kantar, M., Tomlinson, B., & Raghavan, B. How Sustainable Is the Smart Farm? In Proceedings of the LIMITS Workshop on Computing within Limits, Virtual, 14–15 June 2021. (Google Scholar) (CrossRef)

Styawati, S. (2023). Mustard Plant Growth Monitoring System Using Artificial Intelligence in Aquaponics. *IC-ITECHS*, *4*(1), 37–44.

Taha, M. F., ElMasry, G., Gouda, M., Zhou, L., Liang, N., Abdalla, A., Rousseau, D., & Qiu, Z. (2022). Recent advances of smart systems and internet of things (iot) for aquaponics automation: A comprehensive overview. *Chemosensors (Basel, Switzerland)*, *10*(8), 303. DOI: 10.3390/chemosensors10080303

Tik, L. B., Khuan, C. T., & Palaniappan, S. (2009). Monitoring of an aeroponic greenhouse with a sensor network, IJCSNS. *Inter J of Compu Sci and Net Sec.*, *9*(3), 240–246.

Tirlangi, S., Teotia, S., Padmapriya, G., Senthil Kumar, S., Dhotre, S., & Boopathi, S. (2024). Cloud Computing and Machine Learning in the Green Power Sector: Data Management and Analysis for Sustainable Energy. In *Developments Towards Next Generation Intelligent Systems for Sustainable Development* (pp. 148–179). IGI Global. DOI: 10.4018/979-8-3693-5643-2.ch006

Torres-Tello, J., & Ko, S.-B. (2023). Interpretability of artificial intelligence models that use data fusion to predict yield in aeroponics. *Journal of Ambient Intelligence and Humanized Computing*, *14*(4), 3331–3342. DOI: 10.1007/s12652-021-03470-9

Tzounis, A., Katsoulas, N., Bartzanas, T., & Kittas, C. (2017). Internet of Things in agriculture, recent advances and future challenges. *Biosystems Engineering, 164*, 31–48. DOI: 10.1016/j.biosystemseng.2017.09.007

Vangeri, A. K., Bathrinath, S., Anand, M. C. J., Shanmugathai, M., Meenatchi, N., & Boopathi, S. (2024). Green Supply Chain Management in Eco-Friendly Sustainable Manufacturing Industries. In *Environmental Applications of Carbon-Based Materials* (pp. 253–287). IGI Global., DOI: 10.4018/979-8-3693-3625-0.ch010

Venkateswaran, N., Kiran Kumar, K., Maheswari, K., Kumar Reddy, R. V., & Boopathi, S. (2024). Optimizing IoT Data Aggregation: Hybrid Firefly-Artificial Bee Colony Algorithm for Enhanced Efficiency in Agriculture. *AGRIS On-Line Papers in Economics and Informatics, 16*(1), 117–130. DOI: 10.7160/aol.2024.160110

Vennila, T., Karuna, M., Srivastava, B. K., Venugopal, J., Surakasi, R., & Sampath, B. (2022). New Strategies in Treatment and Enzymatic Processes: Ethanol Production From Sugarcane Bagasse. In *Human Agro-Energy Optimization for Business and Industry* (pp. 219–240). IGI Global.

Verma, R., Christiana, M. B. V., Maheswari, M., Srinivasan, V., Patro, P., Dari, S. S., & Boopathi, S. (2024). Intelligent Physarum Solver for Profit Maximization in Oligopolistic Supply Chain Networks. In *AI and Machine Learning Impacts in Intelligent Supply Chain* (pp. 156–179). IGI Global. DOI: 10.4018/979-8-3693-1347-3.ch011

Wang, M., & Xu, L. (2022). Efficiency improvements in aeroponic nutrient delivery systems: A review. *Journal of Agricultural Engineering Research, 134*, 95–105. DOI: 10.1016/j.jaer.2022.01.014

Williams, L., & Davis, K. (2024). Enhancing crop yield through advanced aeroponic systems: A case study. *Vertical Farming Journal, 10*(2), 45–60.

Wittmann, S., Jüttner, I., & Mempel, H. (2020). Indoor Farming Marjoram Production—Quality, Resource Efficiency, and Potential of Application. [Google Scholar] [CrossRef]. *Agronomy (Basel), 10*(11), 1769. DOI: 10.3390/agronomy10111769

Xu, J., Gu, B., & Tian, G. (2022). Review of agricultural IoT technology. *Artificial Intelligence in Agriculture*, *6*, 10–22. DOI: 10.1016/j.aiia.2022.01.001

Yépez-Ponce, D. F., Salcedo, J. V., Rosero-Montalvo, P. D., & Sanchis, J. (2023). Mobile Robotics in Smart Farming: Current Trends and Applications. [Google Scholar] [CrossRef] [PubMed]. *Frontiers in Artificial Intelligence*, *6*, 1213330. DOI: 10.3389/frai.2023.1213330 PMID: 37719082

Zekrifa, D. M. S., Kulkarni, M., Bhagyalakshmi, A., Devireddy, N., Gupta, S., & Boopathi, S. (2023). Integrating Machine Learning and AI for Improved Hydrological Modeling and Water Resource Management. In *Artificial Intelligence Applications in Water Treatment and Water Resource Management* (pp. 46–70). IGI Global. DOI: 10.4018/978-1-6684-6791-6.ch003

Zhang, H., & Yu, Z. (2024). The future of urban agriculture: Combining aeroponics with renewable energy sources. *Journal of Sustainable Agriculture*, *32*(3), 120–133. DOI: 10.1007/s11628-024-00712-3

About the Contributors

G. Chelladurai is currently an Assistant Professor of Zoology at St. Joseph's College (Autonomous), Tiruchirappalli, Tamil Nadu, India. During his doctoral studies, he worked on a project with the Indian National Centre for Ocean Information Science, Government of India, focused on monitoring and locating fisheries-rich areas and a satellite-mediated tsunami warning system using a wave rider buoy system in southeast coast zone. He is an accomplished researcher in coastal aquaculture, with a particular focus on marine gastropods and immunomodulation. He has published a total of 70 research papers in prestigious journals, many of which are indexed in SCOPUS and Web of Science. He has also published 18 books with ISBNs and 28 book chapters. Dr. Cheladurai has reported the occurrence of two new fish ectoparasites on Gulf of Mannar region and registered several Indian patents in biotechnology and aquaculture field, contributing to innovations in sustainability and health science. He has been recognized with multiple awards and is actively engaged in community programs promoting intellectual property rights and environmental safety. He was awarded two research grants from the TNSCST, NCST as well as five additional research grants from government and private organizations for program implementation. Recently, he received the Swachhta Saarthi Fellowship (Cohort 2023) from the Department of Science and Technology, Government of India, and an Innovative Voucher Programme Grant from the Entrepreneurship Development Cell, Government of Tamil Nadu.

***

Vandita is a Ph.D. scholar of Forest Ecology & Climate Change Division, Forest Research Institute Deemed University, Dehradun, India. She holds M.Sc. degree in Environmental Sciences from University of Allahabad, India. She also holds UGC NET-JRF award, and her doctoral research is in the field of Forest Ecology. She also has research experience in the field of Atmospheric Sciences and data analysis using RS & GIS

Sampath Boopathi is an accomplished individual with a strong academic background and extensive research experience. He completed his undergraduate studies in Mechanical Engineering and pursued his postgraduate studies in the field of Computer-Aided Design. Dr. Boopathi obtained his Ph.D. from Anna University, focusing his research on Manufacturing and optimization. Throughout his career, Dr. Boopathi has made significant contributions to the field of engineering. He has authored and published over 225 research articles in internationally peer-reviewed journals, highlighting his expertise and dedication to advancing knowledge in his area of specialization. His research output demonstrates his commitment to conducting rigorous and impactful research. In addition to his research publications, Dr. Boopathi has also been granted one patent and has three published patents to his name. This indicates his innovative thinking and ability to develop practical solutions to real-world engineering challenges. With 17 years of academic and research experience, Dr. Boopathi has enriched the engineering community through his teaching and mentorship roles.

Deepika Dhamija is a seasoned academician with 14 years of experience, currently serving as a Senior Assistant Professor in the Department of Computer Applications at the Centre for Distance and Online Education, Manipal University Jaipur. Proficient in blended-teaching methodologies, she has an extensive portfolio of research publications in international peer-reviewed journals and conferences, alongside book chapters with leading international publishers.Her research interests encompass digital payments, security in mobile payments, mobile application testing, cyber-security, cloud computing, educational technology, and e-governance. Dr. Dhamija has contributed to Scopus-indexed book chapters and published an e-book. She has actively participated in numerous FDPs, workshops, and seminars organized by prestigious institutions and researchers. Dr. Dhamija's accolades include the Best Paper Award, Young Scientist Award, and Woman Researcher Award at various conferences and seminars.

Geetha Manoharan is currently working in Telangana as an assistant professor at SR University. She is the university-level PhD program coordinator and has also been given the additional responsibility of In Charge Director of Publications and Patents under the Research Division at SR University.

Frolian Mobo is a Doctor of Public Administration graduate from the Urdaneta City University Class of 2016 and a graduate of the 2nd Doctorate Degree (Ph.D.) in Development Education program at the Central Luzon State University, Nueva Ecija, Philippines, Class of 2022. On March 11, 2024, Dr. Mobo was accredited and reclassified by the Commission on Higher Education (CHED) to the position of Professor II in the Philippine Merchant Marine Academy (PMMA), and this allowed him to work with different international research institutions, such as the Director and Research Consultant of the IKSAD Research Institute, Turkey. At present, he is in the process of finishing his 3rd master's degree, leading to social studies education at Bicol University. Recently, Dr. Mobo passed Batch 3—Certified Research Professional—and ranked in the top 5 in the National Examination.

R. Vettriselvan, an Associate Professor, AMET Business School, AMET University, specialised in HRM and Marketing. He is acting as Mentor, Saraswathi Institute of Medical Sciences, Hapur. He served as Review Board member, National Council for Higher Education, Malawi, Head of the Department, School of Commerce and Management Studies, St. Eugene University, Zambia, and Director, Research and Publication, St. John the Baptist University, Malawi. He has published 20 books and 75 articles in peer reviewed journals and edited book chapters. Under his guidance two have completed and five are pursuing their PhD.

S. Sampath Kumar holds a PhD in Information and Communication Engineering from Anna University in India. He is now an Associate Professor in the Computer Science and Engineering Department at Sri

Eshwar College of Engineering in Coimbatore. His areas of expertise include Deep Learning, Machine Learning, Cloud Computing, Theoretical Computation, and Unmanned Aerial Vehicles. Currently, he is engaged in Cloud Computing and Unmanned Aerial Vehicle projects, showcasing his dedication to the field.

Index

www.ingramcontent.com/pod-product-compliance
Lightning Source LLC
LaVergne TN
LVHW061543290926
849162LV00004B/892